ENERGY MANAGER'S WORKBOOK
VOLUME 2

ENERGY MANAGER'S WORKBOOK

Based on selected contributions and papers
presented to the Energy Managers' Workshops
organised jointly by the British Institute of Management
and the Department of Energy

Volume 2

 **Energy Publications**

First published in Britain in 1985 by Energy Publications

©   The British Institute of Management

ISBN 0 905332 39 3

Energy Publications is the book publishing imprint of
Cambridge Information and Research Services Limited
Grosvenor House, High Street, Newmarket, England

Printed in Great Britain by
St Edmundsbury Press, Suffolk

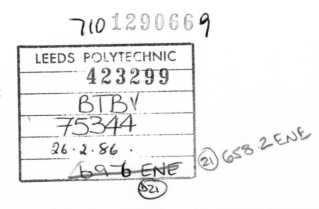

CONTENTS

FOREWORD

by Gerard A D Coghlan OBE, JP
Chairman of West Birmingham Health Authority

Commitment to reducing energy costs, and thereby improving energy efficiency, must come from the top. In industrial and commercial terms this means the Chief Executive of the organisation without whose support the energy manager is deprived of his 'clout'. This is why I have been particularly pleased to be involved with the programme of Energy Managers' Workshops, which are jointly organised by the Department of Energy and the British Institute of Management, and which have formed the basis for the first and, now, this second volume of the Energy Manager's Workbook.

My own experience as an energy manager was with a large group of industrial companies. The Chief Executive instructed me to form an Energy Committee and personally endorsed my terms of reference. This initial support was followed by regular reports on progress to the Board. Indeed it was made clear at the inaugural management conference that all concerned with the Energy Efficiency Programme would enjoy the backing of the Main Group Board.

Of equal importance is the need for energy cost savings to be acknowledged throughout the organisation as a direct contribution to profitability. Success in this field can be measured 'on the bottom line' and the wider these results are circulated, the more enthusiastic, in my experience, becomes the commitment from both management and workforce.

These are two of the principal messages which are returned to regularly at the Energy Managers' Workshops. Over the past three years we have had the pleasure of welcoming hundreds of energy managers and executives from both the private and the public sector. Some have come with considerable experience behind them whilst others, newly appointed, have been seeking an induction course.

The popularity of the first volume of the Workbook, based on the teaching material of these Workshops, has surpassed all expectation and I have no hesitation in commending Volume 2. It incorporates a number of new presentations at the Workshops and some carefully selected additional material dealing with new techniques and approaches to energy management.

Energy managers need to display a high level of resourcefulness and adaptability in addition to their sound technical grounding. Theirs is a job of investigation, analysis, recommendation and persuasion. They need to pull together many disciplines—technical/production, personnel and financial—to achieve the twin objectives of reducing energy costs and improving energy efficiency. The two volumes of the Energy Manager's Workbook will prove an invaluable aid in meeting these objectives.

Gerard Coghlan is the Course Director of the Energy Managers' Workshops on behalf of the Department of Energy and the British Institute of Management.

INTRODUCTION

by Peter Martin
Course Administrator, The Energy Managers' Workshops

The strength of the Energy Managers' Workshops has resided in the preparation and expertise of the team of specialists who have contributed to them over the past three years. What they had to say seemed so good that the message simply had to be conveyed to a wider audience. Hence the decision to produce and publish the first volume of the Energy Manager's Workbook.

Volume 2 is designed to complement the first Workbook. It extends the range of subjects covered by drawing partly on the expertise of speakers and partly on the knowledge of other experts who have not been invited to contribute to the Workshops simply because there is only so much that can be covered within the four-day course. The ten chapters which follow cover topics which have either been added to the Workshops since the first Workbook appeared or have been the subjects of delegate questions and about which more information is now in demand.

The first chapter tackles the key problem facing many energy managers, how to identify investment priorities. The approach flows logically from Barry Healey's chapter in Volume 1 on 'Preparing the Case for Investment' and between them the chapters form a sound basis for identifying and ranking the projects to be tackled. Chapter 2, 'Monitoring Consumption and Setting Targets' takes forward David Yuill's contributions in Volume 1 on carrying out an energy audit. In similar style to Volume 1 this chapter, which deals primarily with the principles and techniques involved, is complemented by a final chapter setting out the experiences encountered when putting the approach into practice. Chapter 10, prepared by a member of the engineering team at Manchester Steel, provides a detailed insight into the results achieved.

Two recurring questions on the Workshops have been firstly how to motivate management and workforce to take energy efficiency seriously, and secondly how to get to grips with electricity charges, selecting the most cost-effective tariff and avoiding load and power factor penalties. These subjects form the bases for Chapter 3 and 4. David Yuill's chapter supports his claim that the energy manager should, in consultation with his local Electricity Board but without the help of an outside consultant, be able to identify the right tariff for his operation and limit load penalties to the minimum.

Chapter 5 'Energy Consumption in Buildings' is taken from one of the most highly regarded sessions of the Workshops. Ken Spiers systematically reviews the many, and sometimes conflicting, influences affecting heating needs. He teaches delegates to look at buildings in a new light and to probe for the improvements which can be made.

Heat recovery offers industry by far the greatest opportunity for improving energy efficiency. Glenn Brookes, a recognised authority in this specialist field, shows how these opportunities can be identified, where to use the heat recovered, how to assess the financial case and concludes with a survey of the range of equipment now on the market. Chapter 7 follows on to describe the exciting new technique of process integration which enables energy managers to identify the minimum level of energy required to meet plant needs and maximise the use of waste heat.

Energy management technologies continue to respond to high fuel costs. A feature of recent Workshops has been the session, taken by David Wragg, reviewing these developments and relating them to applications within factories and offices. In Chapter 8 he reviews four developments: CHP generation, heat pumps, induction motor controls and carbon dioxide monitoring. Finally Alan Tweedale summarises the opportunities and benefits now being provided by heat management companies. His chapter takes a look at the range of services provided and the methods for charging now being adopted for those who wish to delegate responsibility for their energy needs.

NOTES ON CONTRIBUTORS

Philip Gardner is a Principal Scientific Officer at the Nuclear Physics Division of AERE Harwell, where he is concerned with the business development of ion beam techniques. He was previously Project Officer at the Energy Technology Support Unit at Harwell, when he was involved with the management of government-supported independently-monitored demonstrations of electronic energy management systems. He lectures frequently on energy management system demonstrations, and on energy efficiency investment appraisal. Author of the book 'Energy Management Systems in Buildings—The Practical Lessions', Dr Gardner has also written a number of articles on energy management technology.

Peter Harris is the Director of the Energy Information Centre. He was involved in the planning of the Department of Energy's early programmes in energy conservation under the Energy Audit Scheme, and also worked on the Energy Efficiency Demonstration Scheme. From 1974 until 1982 he was a member of the Energy Technology Support Unit at Harwell, concerned with improving energy use in a variety of industrial sectors. He subsequently became Director of Services for the Energy Users Research Association. He is a frequent speaker on energy topics at courses and conferences and has been involved in training programmes for engineers both in the UK and overseas. Dr Harris is the author of three of the Energy Audit Series reports and has written numerous other papers and articles on energy saving. He is a Chartered Chemist and Member of the Royal Institute of Chemistry.

Peter Martin is a Director of the Fivewell Group which specialises in management development and training, and is also a consultant to the British Institute of Management. He has worked as an associate with the PE, Metra and Inbucon consulting groups. He is responsible for the development and organisation of the Energy Managers' Workshops programme which he has administered for the past four years, on behalf of the British Institute of Management and the Department of Energy.

David Yuill is an independent energy consultant. After qualifying as a Fuel Technologist at Imperial College, London he spent many years in manufacturing industry with general management responsibility up to Managing Director level. In addition to his practical consultancy work he writes and lectures regularly on energy management, conducts on-site training of energy managers, and was a consultant for the Department of Energy's film 'Saver Approach'. He is also involved in the development of new insulation and condensation control techniques. In recognition of his services in the field of energy management he was awarded the MBE in 1984. Mr Yuill is the author of the widely circulated management guide, 'How to Improve Energy Efficiency', published in 1984. He is a Chartered Engineer and Member of the Institute of Energy.

Ken Spiers holds the position of Manager of the Energy Management Group at the Laing Design and Development Centre. For twelve years he worked as an engineer for the National Industrial Fuel Efficiency Service, based in Cambridge and Nottingham, during which time he was responsible for carrying out numerous energy conservation surveys. He subsequently went on to work as Fuel Efficiency Officer with Essex County Council before joining Laing Design and Development, where he continued to conduct energy surveys and to investigate into energy matters in building design and performance. Mr Spiers lectures regularly on building design, heat pump systems and energy management systems. He is a Member of the Institute of Energy.

Glenn Brookes worked for over three years at the Energy Technology Support Unit at Harwell before becoming the Executive Director of the Energy Systems Trade Association. He has carried out extensive market research into the potential for energy efficiency technologies in a range of industries. His work involves him in numerous lectures, workshops, seminars, conferences and exhibitions, and his publications to date include the Journal of Heat Recovery Systems and the ESTA Yearbook 1984. Dr Brookes is a Member of the Royal Society of Chemistry.

Greg Ashton is a consultant on the techniques of process integration. He worked for sometime as Process Design Section Manager for the chemical engineering contracting company, Petrocarbon Developments Limited and is now a member of the Linnhoff March energy consultancy which specialises in the application of Linnhoff Pinch Technology. It is a joint venture company between Professor Bodo Linnhoff and the March Consulting Group. Dr Ashton is a Chartered Engineer and Member of the Institute of Chemical Engineers.

David Wragg is Principal Engineer at the National Industrial Fuel Efficiency Service (NIFES), where he has been responsible for the co-ordination of training courses for energy managers and plant operatives. He is also Head of the Energy Advisory Section for the NIFES East Midlands Region, which tackles problem areas in industrial and commercial energy management. He has a special interest in the NIFES project demonstrating monitoring and targeting techniques in the Drop Forge industry, and in the application of new technologies for saving energy. Mr Wragg is a Member of the Institution of Mechanical Engineers.

Alan Tweedale was until recently Managing Director of the heat management company, Associated Heat Services PLC, and is now Managing Director of Energy Components Ltd. He has been closely involved with the Department of Energy and with the Energy Technology Support Unit in the development of the use of waste as a fuel source. He lectures regularly on energy matters and is co-author of a book which is currently being written on the potential of waste as a fuel source. Mr Tweedale is a Chartered Engineer and is a member of both the Institute of Gas Engineers and the Institute of Energy.

Stan Smith is the Chief Engineer of Manchester Steel Ltd. He has been directly involved in the introduction and successful implementation of an energy monitoring and target setting scheme. His contribution in Chapter 10 underlines the very real benefits which can accrue from adopting this approach to cutting energy costs.

ILLUSTRATIONS

IDENTIFYING INVESTMENT PRIORITIES

Philip Gardner

Energy managers and others with responsibility for maintaining and improving energy efficiency face many difficulties. One particular problem is obtaining finance for energy efficiency improvement projects. A common complaint is that money is simply not available for this type of investment.

Part of the reason lies in the weakness of the cases that are presented to the financial decision-makers. Energy managers normally have limited spending authority; their job is to make technical judgements and to advise more senior managers on how to spend money in their specialised area. The senior manager's job is to take the broader view and to consider the organisation's overall interests. In doing this, he will receive advice from middle managers in other specialised areas such as production, research and marketing.

Unfortunately for the energy manager, his case often fails to be sufficiently convincing to compete successfully with the many other worthy claims on scarce capital resources. He will find himself at a particular disadvantage if he fails to communicate well with the finance executive, who may have little technical understanding.

This chapter is intended to help energy managers to pinpoint the energy efficiency investments most likely to produce an attractive rate of return, and to present financial cases that maximise the prospects of eventual approval. This might result in more money being channelled into energy efficiency improvement and perhaps less into production. However, if this increases the profitability and competitiveness of the organisation and helps to secure its long-term future, then this might be the right course of action.

Some of the points raised here are dealt with in more detail in Volume 1 of the Workbook and elsewhere in this volume. This chapter seeks to bring together several important aspects of an energy manager's job and to focus on a set of actions aimed at cost-effective reductions in energy bills.

Preparing the Ground

High energy efficiency, or any other sort of efficiency for that matter, will be difficult to achieve unless an important set of organisational and managerial circumstances prevails. It is not in the power of the energy manager to bring about all of these, but he will wield some influence in certain areas.

Involving the Financial Decision-Makers

The first requirement is that the financial decision-makers are fully alerted to the real benefits of energy efficiency improvement. Energy costs may be only a small proportion of an organisation's annual turnover but probably represent a significant proportion of total profits.

Senior managers need to realise that reductions in operating costs—such as for energy—can add directly to profitability. If individual investments in energy efficiency improvement offer acceptable rates of return and are within the company's ability to undertake and operate, then it is difficult to argue against their implementation. Unfortunately, not all senior managers appreciate this point, in many cases concentration is understandably focused on the major issues facing the organisation's development. Unless they are receptive to the energy manager's objectives, he is unlikely to achieve them.

The Need for Good Projects and Proposals

There is little value in switching senior management onto the prospects for energy efficiency investment if good proposals do not come forth promptly from middle management. Energy managers should be able to: identify all the appropriate opportunities; provide thorough technical and financial assessments; pick out the options representing best value for money; and present well-argued and credible cases for investment.

This is where many energy managers falter: they sometimes fail to recognise valuable cost reduction opportunities; inappropriate projects are put forward; the financial assessments can be crude to the point of being meaningless; and even good investment opportunities will be passed over by the Board if proposals are presented poorly and the case for support argued weakly.

Furthermore, the body of the organisation—management and employees generally—need to be receptive to the notion of improving energy efficiency. This is particularly important where buildings and comfort conditions are concerned. If people have got used to excessive comfort conditions, it can be extremely difficult to reduce costs by cutting back on space temperatures.

It may also prove difficult to eliminate unnecessary use of energy—e.g. for lighting—if the climate of opinion is unfavourable. Even improvements to process efficiency can invoke hostility if they are implemented in an insensitive way. It is vital that employees appreciate what is being done and why; they also need to understand and accept the underlying importance of maximising the efficient use of all resources, not only of energy but of manpower, materials and money. Education and training have a clear role to play here, and this aspect is discussed separately later in this chapter.

Getting the Basics Done First

Finally, there is no point in embarking on a major programme of energy efficiency capital investment if nothing has been done previously to bring the existing plant, processes and buildings up to maximum achievable standards through revenue expenditure measures.

Existing plant should have received the benefits of proper operations and maintenance before consideration is given to adding new bits of hardware such as heat recovery or advanced monitoring and control systems. Otherwise, it is like putting a sophisticated electronic engine control system on a clapped-out, badly maintained, gas-guzzling old car and expecting miraculous improvements in fuel economy. Unless you know how well your existing plant can perform, it is hard to make out a convincing case for spending large sums of money on upgrading or replacement.

Tackling the Barriers

Even when propitious organisational circumstances prevail, the ambitious and eager energy manager still has many problems to tackle before he can improve energy efficiency. Some of these problems stem from the very real barriers to investment in energy efficiency that face most organisations. Energy managers need to recognise and understand these barriers if they are to overcome them.

Lack of Awareness

Many energy efficiency improvement measures fail to be implemented because the prospective user is unaware of their existence or of the likely obtainable benefits. An important example is lighting control. Inefficient tungsten filament lamps are being replaced in many industrial buildings by high-efficiency high pressure sodium units. However, unnecessary use of the new lamps for long periods when the factory is unoccupied continues as before because of inadequate manual switching arrangements.

The potential of a lighting control system which automatically switches lights out at strategic times is often overlooked. This may be because the energy manager is unaware of the technology or mistakenly believes it is unlikely to be cost-effective or acceptable to the workers. All appropriate measures need to be considered and assessed.

An Energy Efficiency Office has been established at the Department of Energy in an attempt to improve good awareness of the opportunities for cutting energy costs. It includes a team of Regional Energy Efficiency Officers from whom information and advice can be obtained—see Table 1.1.

Competing with Other Demands for Funds

Most organisations have a far more relaxed attitude about investment in their mainstream activities than they have about energy efficiency investment. A payback of up to ten years will be accepted for investment in production, whereas returns of capital within one year are often demanded for energy efficiency improvement.

To some extent this is understandable. A business is not run on energy saving; a manufacturer must keep ahead of the competition in design, performance and price or go out of business. On the other hand, energy cost reduction can provide a valuable contribution, for which carefully selected, assessed and well-argued proposals are essential. The problem underlines the importance of having a receptive senior management and being able to communicate well with financial people. This chapter is intended to help in these areas.

Lack of Information about Specific Measures

Even when an organisation is aware of a range of appropriate energy efficiency improvement measures, it may still desist from investing because of insufficient information about the techniques involved. This may be especially true with new technology. The Energy Efficiency Office is active here also with its Energy Efficiency Demonstration Scheme (EEDS). Additional valuable information can be obtained from existing users. If they are involved in the EEDS, they are obliged to accommodate enquiries, but even if they are not, many will be pleased to help if no breach of commercial confidence is involved. Contact with the local Regional Energy Efficiency Officer is a good first step to take.

Perceived Risk of Failure

Many organisations hold back from embarking on energy saving projects because they fear that things may go wrong and more money may be wasted than saved. This is especially true with techniques outside their areas of expertise or experience. Such fears are understandable but often exaggerated. Much can be done to allay them by making full use of available information and existing users' experiences.

Lack of Required Expertise

Projects may not go ahead because there are no suitably qualified and able staff to implement and run them. Consultants can play a valuable role in the initial stages of selecting and assessing opportunities, specifying equipment, assessing tenders, installation and commissioning, acceptance

TABLE 1.1: THE REGIONAL ENERGY EFFICIENCY OFFICES

SOUTH WEST REGION

Gloucestershire, Avon, Wiltshire, Somerset, Dorset, Devon, Cornwall (incl Isles of Scilly)

Doug Ponsford
The Pithay
Bristol BS1 2PB
Tel: (0272) 291071 ext 205
Telex: 44214

SOUTH EAST REGION

Dr Alan Franklin
Tel: 01-603 2060 ext 246

(i) Southern Area

Greater London, Surrey, Kent, East Sussex, West Sussex, Hants, Isle of Wight

Maurice Webb
Charles House
375 Kensington High Street
London W14 8QH
Tel: 01-603 2060 ext 245
Telex: 25991

(ii) Eastern Area

Norfolk, Suffolk, Cambridgeshire, Bedfordshire, Buckinghamshire, Oxfordshire, Berkshire, Hertfordshire, Essex

Godfrey Smith
Charles House
375 Kensington High Street
London W14 8QH
Tel: 01-603 2060 ext 248
Telex: 25991

WALES

Gerry Madden
Industry Department
Cathays Park
Cardiff CF1 1NQ
Tel: (0222) 823126
Telex: 498228

WEST MIDLANDS REGION

Staffordshire, Shropshire, West Midlands, Warwickshire, Hereford & Worcester

Bob Anthony
Ladywood House
Stephenson Street
Birmingham B2 4DT
Tel: 021-632 4111 ext 549
Telex: 337021

EAST MIDLANDS REGION

Derbyshire (except High Peak District), Nottinghamshire, Lincolnshire, Leicestershire, Northamptonshire

David Warren
Severns House
20 Middle Pavement
Nottingham NG1 7DW
Tel: (0602) 56181 ext 285-4-3
Telex: 37143

YORKSHIRE & HUMBERSIDE REGION

North Yorkshire, West Yorkshire, South Yorkshire, Humberside

Jim Gilroy
Priestley House
Park Row
Leeds LS1 5LF
Tel: (0532) 443171 ext 213
Telex: 557925

NORTH WEST REGION

Cumbria, Lancashire, Merseyside, Greater Manchester, Cheshire, Derbyshire High Peak District

Robin Gardner
Sunley Building
Piccadilly Plaza
Manchester M1 4BA
Tel: 061-236 2171 ext 640
Telex: 557925

NORTH EAST REGION

Northumberland, Tyne and Wear, Durham, Cleveland

Arthur Hoare
Stanegate House
Newcastle Upon Tyne NE1 1YN
Tel: (0632) 324722 ext 217
Telex: 5317

NORTHERN IRELAND

Derek Noble
Department of Commerce
Chichester House
64 Chichester Street
Belfast BT1 4JX
Tel: (0232) 234488 ext 420
Telex: 747152

SCOTLAND

Ian Melville
Industry Department for Scotland
Energy Division R6/54
New St Andrews House
St James Centre
Edinburgh EH1 3SX
Tel: 031-556 8400 ext 4572
Telex: 727301

Source: Department of Energy

testing and utilisation. Some may take on the long-term running of a project on behalf of the client—e.g. bureau services for electronic energy management systems.

However, most users will eventually want to take over direct responsibility for their projects. Specialist technical training is obviously very important here.

Problems with Buildings

There are special problems with improving energy efficiency in buildings. Comfort conditions may be affected and occupants alienated. If the building is tenanted, there is the difficulty of the capital costs falling on the landlord while the tenants reap the benefits.

The poor standards of design and construction of much of our building fabric and services not only add to energy costs, but make it especially difficult to implement improvement measures. The long life-spans of buildings mean that many older ones have obsolete plant and inadequate controls. Sometimes it is better to demolish and replace an old building rather than try to implement a series of retrofit energy efficiency improvement measures.

Uncertainties Over the Energy Future

Changes in the real price of energy are an important factor in the financial appraisal of energy efficiency investment opportunities. Organisations are more keen to invest when they see the prices of fuel going up faster than average inflation. For nearly a decade after the huge oil price increases of the early 1970s, it was assumed that the real price of energy would continue to climb indefinitely.

Recently, it has been seen that this is not necessarily the case; the real prices of oil and coal have been declining since 1982 because of world overproduction during recession. This has undermined the confidence of some prospective investors. However, it is reasonable to assume that the longer-term outlook is for cheap sources of energy to be in increasingly short supply and real prices to rise steadily. Energy managers need to be alert to energy market trends and to be able to give sound and convincing advice to their senior colleagues.

Making the Most of the Revenue Account

A sensible energy efficiency improvement strategy will exploit the full potential of revenue expenditure before measures involving capital expenditure are embarked upon. For one reason, it is usually easier to obtain financial approval for revenue items. For another, the financial benefits can be fed back into the energy budget to fund further improvements. Furthermore, such a strategy is likely to produce the right organisational and technical circumstances to ensure the success of subsequent capital projects.

Energy managers should strive to make the most of three broad types of measure:

—improvements in operations and maintenance
—education and training
—shaping administrative policy.

Each measure is now discussed in turn.

Improvements in Operations and Maintenance

As mentioned previously, it makes good sense to raise the standards of operations and maintenance of existing plant to the highest practical levels before investing in new plant or retrofit efficiency improvements. For example, boilers, controls and heat distribution systems should be made to run as efficiently as possible before installing an expensive electronic energy management system.

The new monitoring and control system will then stand a reasonable chance of working effectively. Moreover, it will help to create the right levels of satisfaction among building occupants; they might otherwise react antagonistically to measures which they perceive to be intended to erode their already unsatisfactory comfort conditions.

Education and Training

A well thought-out publicity campaign should convince all staff of the need for good standards of housekeeping and energy awareness. They should appreciate that it is in their best interests that all unnecessary and excessive use of energy be eliminated.

Energy cost savings add directly to profit. They will help safeguard employees' futures by improving the firm's economic well-being and competitiveness. Moreover, each pound saved is equivalent to many pounds' worth of extra production. Energy efficiency is actually an easy way to increase the earnings of the firm! It is important to emphasise that sacrifices are not being sought, nor are the staff being expected to work in less than satisfactory conditions. A good example is minimising the opening of windows as a means of room temperature control in winter—or even in summer for air-conditioned buildings.

Staff should be encouraged to use the radiator valve or room thermostat to regulate temperature. Lights should be switched out where they are no longer required and care should be taken with the use of hot water. Significant benefits can be obtained from this type of campaign, if it is handled properly.

Early encouraging results are unlikely to be sustained indefinitely. People do tend to drift back into their former habits, but the right climate of opinion will be established for introducing more sophisticated, expensive and lasting measures at a later date.

Shaping Administrative Policy to Energy Costs

A major advance towards higher energy efficiency will result from an adjustment of the administration and policy to take into account the importance of energy costs within total overheads. This might start by identifying these costs—many organisations do not know how much they spend on energy because the figures are lumped together with several other costs such as water charges. The energy costs need to be broken down according to fuel type and, as far as possible, end use. This will be dealt with in more detail in the following section.

The next stage might be the allocation of responsibility for energy management and a firm budget for the costs. If the annual energy bills are sufficiently large, the appointment of a full-time energy manager might be justified. Otherwise, the responsibility will probably have to be part of the routine duties of existing technical staff, such as the works or estates manager. In either case, adequate training needs to be given and appropriate authority vested.

It is also worth considering what should be done with the energy cost savings. If the benefits are simply swallowed up into some other more general budget, there is little incentive to maintain or increase them. It is far better to allow at least some of the savings to be invested back into further cost reduction measures. This type of approach can be particularly effective when the time comes to consider capital investments.

Identifying Capital Investment Opportunities

The best way to identify these opportunities in energy efficiency improvement is through a detailed and comprehensive survey of buildings and process plant. Moreover, without a clear idea of how much energy is being used, where it is going and how it is utilised, it will not be possible to derive credible forecasts of the likely financial returns from individual improvement measures.

Lack of in-house expertise and available manpower might be a problem here. In such cases, the necessary expertise can be brought in from one of the many consultancy firms now operating in this field. Some are small, perhaps one-man enterprises specialising in particular aspects such as buildings or process plant. Others are large well-established firms who have expanded into energy efficiency improvement from other mainstream activities, e.g. W S Atkins, Ove Arup, Ewbank Preece, PA Management Consultants and the National Industrial Fuel Efficiency Service. Furthermore, grants towards the costs of such consultancy services can be obtained in approved cases from the Energy Efficiency Office.

However, use of consultants will not relieve the customer of all burdens: he will need to evaluate credentials carefully and to brief and manage the appointee; the final recommendations will need thorough consideration for translation into good proposals for senior management. The consultant's role is to provide the particular expertise that the customer may lack; there can be no substitute for the customer's knowledge of his own plant and buildings and he will still be responsible for any judgements and decisions that are made.

The recommended procedure for conducting an energy survey is given in some detail in Chapter 2 of the Energy Manager's Workbook, Volume 1. Accordingly, only a few guidelines are given here for use either in a consultant's brief or in tackling the programme alone.

Getting to Know the Site

A great deal of valuable qualitative information can be obtained from walking around the buildings simply using eyes and commonsense. Many instances of wasteful or unnecessary uses of energy will soon become apparent.

More detailed quantitative surveys will be required to assess whether any cost-effective measures are available to reduce or eliminate inefficiencies discovered. Some problems might be best tackled using the kinds of measures described in the previous section. Others will require capital expenditure. In either case, there will be no firm basis for decision-making if a good knowledge of the existing situation has not been achieved.

Quantifying Energy Costs

In order to calculate the likely financial return from a specific efficiency improvement investment, the actual energy costs of the process or service involved will need to be determined. In many cases this will not be known in advance; most organisations can identify how much they spend on boiler fuel and on electricity, but few know the costs of items such as domestic hot water or lighting. Such costs will need to be measured or reliably estimated before investment proposals can be constructed.

Some of the necessary information can come only from specially installed sub-meters, such as for electricity on lighting circuits or for steam to DHW calorifiers.

This instrumentation is not cheap to install or to read at regular intervals, so carefully balanced judgements need to be made about its use. Certainly for large sites or energy intensive processes, comprehensive sub-metering can pay dividends in providing the basis for investment appraisal and in improving the overall base of energy management.

The breakdown of energy costs might be in terms of fuel type, e.g. into divisions of oil, gas, electricity and coal. This should then be extended as far as practical and economic to end uses, such as:

—heating
—lighting
—domestic hot water
—manufacturing processes
—services (e.g. compressed air).

Calculating Energy Flows and Identifying Areas of Major Saving

The third step is to focus attention on the individual end uses of energy, using the techniques described by David Yuill in Volume 1 of the Workbook. This might involve the use of so-called Sankey diagrams to determine:

— energy inputs (fuel)
— energy outputs (heat, light, power, etc)
— losses (e.g. in hot waste gases, liquids or solids discharged to the environment).

Examination of the survey data should reveal the major opportunities for reducing energy costs. These might include:

— unnecessary use of energy
— excessive heat loss from insufficiently insulated enclosures
— inefficient operation of plant
— recoverable and re-usable heat.

Examples of typical capital investment opportunities are given in Table 1.2. These opportunities are covered in some detail in the other chapters in this volume and Volume 1 of the Workbook.

TABLE 1.2: EXAMPLES OF TYPICAL OPPORTUNITIES FOR INVESTMENT IN ENERGY EFFICIENCY IMPROVEMENT

Defect	*Remedy*
Steam distribution losses	Replacement of central boilerhouse by modular units
Lighting on when not required	Automatic controls
Excessive comfort conditions and heating on outside normal hours	Better heating controls
Re-usable heat	Heat recovery system
Inefficient plant	Replacement by improved designs or better process control
Unused combustible waste	Waste heat boilers
Poor overall standards of monitoring and control	Comprehensive energy management system

Assembling the Likely Runners

The previous section outlined the ways to identify appropriate energy efficiency improvement measures. The next logical step is to gather together for each measure all the necessary financial data for its full and proper assessment. The key point is that the information collected should be accurate and complete; if wrong or incomplete data go into the calculations, the answers will be meaningless, however sophisticated the financial appraisal techniques.

Part of the necessary information at this stage will arise from comprehensive energy surveys. Much of the remainder will come from submissions by equipment suppliers; a well-conducted equipment specification and tendering exercise is invaluable for maximising savings. Two important points are detailed below.

Identifying All Likely Costs and Benefits

It is vital that all the likely costs and savings expected from each energy efficiency improvement should be quantified. A common mistake is to concentrate just on capital and installation costs and on the energy cost savings. These will certainly be the most important items, but there will be others. To omit them may give a false estimate of the likely rate of return on the investment and possibly cause the best opportunity to be passed over in favour of an inferior one.

A more complete list of costs and benefits for each opportunity is offered in Table 1.3.

TABLE 1.3: THE COSTS AND BENEFITS TO BE CONSIDERED WITH PLANNING ENERGY EFFICIENCY INVESTMENTS

Costs	Benefits
Capital cost of equipment	Energy cost reductions
Engineering costs (to enable equipment to be installed)	Manpower savings
Installation costs	Water savings
Maintenance	Waste disposal cost savings
Operating costs (e.g. telephone charges for an energy management system)	Savings in materials
Staff costs	

Predicting Timescales

The timing of costs and benefits is important in investment analysis because of the change in the real value of money that would take place as a function of time, even if zero inflation was ever achieved. The reason for this change is straightforward: anyone offered the choice of £100 today or in a year's time would obviously choose the former; the £100 could be invested in a building society and turned into perhaps £105 in twelve months. The opposite is true of debts; everyone likes to delay their payment for as long as possible.

Important cost-incurring milestones in a typical prospective project might be:

—feasibility studies
—preliminary engineering
—installation
—commissioning and acceptance
—utilisation

Estimates of the timing of the major project milestones must be realistic. Most items take longer to complete than is initially estimated, especially where some technical development is required. It is prudent to be sceptical of equipment suppliers' forecasts of likely rates of progress and to allow for substantial delays in the programme.

Picking the Winners

Having assembled a set of possible investment opportunities in energy efficiency improvement, the energy manager might have one of three objectives. Firstly, he may be seeking to pick out all those opportunities that meet certain minimum targets on rates of financial return, in order that they may all be implemented in due course. Alternatively, if capital resources are scarce, he might wish to recommend a single investment opportunity offering the highest rate of return. Finally, his aim might be to exploit to the full a finite capital budget, by selecting a group of projects that make best use of the available funds, and perhaps to leave as little as possible unspent at the end of the financial year. In all these cases, rigorous financial appraisal is essential if value for money is to be maximised.

Simple Payback

Most energy efficiency investment appraisal extends no further than calculating the expected payback time, i.e. capital costs divided by the expected annual cost savings. This is a good measure of the likely acceptability of a project; no amount of more sophisticated analysis will turn an anticipated 30 year payback time into a golden investment opportunity.

Simple payback analysis is useful to eliminate the no-hopers, but will not show which project represents the best opportunity out of a number of options. It cannot do this because it fails to take into account:

—the timing of costs and benefits
—the likely residual value of assets at the end of the project life (they could be sold)
—savings accruing after the payback term (these may increase with time).

For example, simple payback cannot tell which of two opportunities, both with the same payback period, represents the better investment. Nor can it confirm that a prospective one-year payback is as good an opportunity as it may seem. Such a rapid return is often quoted for difficult and demanding measures such as electronic energy management systems. In fact, these can take as much as a year to install and commission and a further two years for the purchaser to learn how to use them properly. Under these circumstances, simple payback periods are relatively meaningless.

To establish the actual rate of return from an investment and to select the best opportunity, a more applied approach may need to be taken.

More Rigorous Financial Appraisal

Techniques of investment appraisal are described in detail by J B Healey in Chapter 3 of Volume 1 of the Workbook. These include discounted cash flow, internal rates of return and net present value of a project. All these techniques take into account the way that the opportunity value of money changes with time. They do this by discounting costs and benefits from the time they are expected to occur back to a common project start date. The discount rate used depends on factors such as prevailing inflation and interest rates and the project risk, but normally lies in the range of between five and ten per cent.

These more rigorous appraisal techniques allow the various investment opportunities under consideration to be rank-ordered. They thereby enable the best opportunities to be selected or the available capital budget to be used to maximum efficiency. They are also the energy manager's only real assurance of value for money.

Weighing Up the Uncertainties

The results from any financial appraisal can never be absolutely precise because the input data contains uncertainties. Project progress may vary from the target dates; the capital costs might escalate; the cost savings might fall within a range of values; the likely operating costs may be difficult to assess accurately.

Consequently, the principal input parameters should be varied and a range of possible rates of return derived for each investment. The best and worst possible cases should be included, together with the most probable value. Sensitivity analysis will enhance the credibility of the investment proposals for the financial decision-makers, by giving them a much fuller idea of what to expect.

Assessing Other Factors

The final aspect of investment appraisal is to consider the non-financial factors; how will the proposed investment fit into the organisation and the way it conducts its business? Will full and proper use really be made of the equipment? For example, is the necessary expertise and manpower available to operate a sophisticated installation like an electronic energy management system? With a heat recovery system, can the waste energy really be put to good use—either back into the process or for other purposes such as space heating—without incurring substantial technical risks?

Presenting the Results and Maintaining Momentum

Bringing together all the points covered so far in this chapter suggests that a more rigorous approach to ensuring best value for money in energy efficiency investment might include the steps outlined in Table 1.4.

TABLE 1.4: THE RIGOROUS APPROACH TO ENERGY EFFICIENCY INVESTMENT

* Define objectives

* Consider all appropriate options

* Identify all costs and benefits

* Discount money values

* Weigh up the uncertainties

* Assess other factors

* Present the results

It is, however, of little value to adopt a rigorous approach to energy efficiency investment unless the details and results can be presented in a clear and convincing way to the people making the final decisions.

A common difficulty concerns assembling the points and ideas in a logical sequence. The structure outlined in Table 1.5 might prove helpful in such cases.

It is important to keep in mind what it is hoped that the proposals will achieve and to make a clear set of recommendations based on carefully constructed arguments. In this way, the proposals are presented in the common language of senior decision-makers, including financial experts, and are more likely to be given proper consideration.

TABLE 1.5: POSSIBLE STRUCTURE OF AN ENERGY EFFICIENCY INVESTMENT PROPOSAL

1. Objectives

2. Options

3. Capital and other costs (undiscounted)

4. Benefits (undiscounted)

5. Rigorous financial appraisal

6. Uncertainties and sensitivities

7. Non-financial factors

8. Option(s) representing best value for money

9. Comparison with important alternatives

10. Recommended course(s) of action

If the initial recommendations prove sound, credibility will be established and the potential contribution of energy efficiency to profitability will begin to be recognised. However, the energy manager should not rest on his laurels after a few initial successes, but rather exploit his advantage to the full.

There will always be scope for considering new investment opportunities. New technologies will be continually introduced. The rising real price of fuel will gradually turn an increasing number of previously uneconomic measures into cost-effective opportunities. Continuous monitoring, combined with rigorous assessment and appraisal as these opportunities arise, forms the key to achieving and maintaining peak energy efficiency.

2
MONITORING CONSUMPTION AND SETTING TARGETS

Peter Harris

It is important that an energy management programme has objectives and a means of monitoring progress towards them. We should want to know that actions we have taken are achieving the expected result and, also, when corrective action is appropriate and in what form. To achieve this we must collect information on the activity of the business and the amount of energy we use, and be able to interpret it correctly. The techniques to do this come under the collective heading of 'monitoring and target setting.'

Proper interpretation is a key part of this activity. Energy use in an industrial process is expected to depend on the quantity of goods produced whilst energy use in a building is expected to depend on weather conditions. There are several approaches that can be taken. However, in practice, the savings achieved from any one measure may be only a small proportion of total energy use and total savings may be due to a number of measures implemented at different times. It would be useful then to be able to evaluate the separate contributions from each and to ensure that each is still giving the savings expected.

To do this successfully requires the information to be handled by very particular methods. The methods are simple enough to use although there is no concise account of their application to energy use which is readily available to the manager or executive. They are therefore covered here in some detail.

Available Information

It is most likely that some of the information needed is already being collected. Production information would be available perhaps as tonnes or litres per week, or bags per shift. Information on energy use might be available at the whole site level, or at process level as therms of gas, kilowatt hours of electricity or pounds of steam. Information on weather would not normally be collected and in any case needs to be of a very particular kind, as discussed later. Usually the information will be available as time-series data: that is, the information is available week by week or month by month from which it should be possible to deduce trends.

The general problem of interpretation is shown by Figure 2.1 which shows production in tonnes per week and energy use in therms per tonne for a process week by week over the course of a year. This is a commonly used form of presentation. We can see that production has been very erratic, although it was reasonably steady in the first three months. Also, specific energy consumption depends on production rate—as production falls, energy consumption per tonne increases.

There is a temptation to regard as the most useful information that which corresponds to periods of steady production. In the example in Figure 2.1 this happens between weeks 1 and 12 when production averaged 71.5 tonnes per week and energy consumption was 0.170 therms per tonne and again between weeks 33 and 41 when production averaged 68.2 tonnes per week and energy consumption averaged 0.156 therms per tonne. On this basis energy consumption seems to be about eight per cent lower in the latter period.

There is some evidence of savings in the third and fourth quarters of the year. The general pattern is that, as production falls, specific energy consumption goes up. However, between the first and the third quarter production was down 16 per cent and energy use down 2.4 per cent. Between the second and fourth quarters production was down 7.8 per cent and energy use fell by 12.5 per cent. The problem we have is that with the information expressed in this form it is difficult to be very much more systematic or precise about the interpretation, and as the information builds up it is difficult to relate new information to what has gone before.

Far more reliable evidence of savings can be extracted from this information but it must be handled rather differently. There are two stages: first to relate existing consumption to production or the weather; and then, using this, to establish time trends.

Relating Energy Use to Production

In relating energy use to production we are likely to be interested in energy use at two levels—the level of energy consumption for the overall plant on site and the level of consumption relating to individual unit processes. Either way the basic method is the same.

Put in general terms, we expect energy use in processes to comprise two parts: a fixed component, known as the **production—unrelated demand**, which continues to function even if we produce nothing, and a variable component, known as the **production—related demand**, which depends on the rate of production. This is exactly analogous to the idea of fixed costs and variable costs in the running of a business. To find out the proportions of each, all that is necessary is to plot a graph of energy consumption against production and extend the line to zero production. (It is useful to plot it out initially marking each point with a date in case there is already some time-related trend in past data to be picked out.)

In the case of the data from Figure 2.1 a consistent pattern emerged for the first 26 weeks, which is shown in Figure 2.2. All the data fit onto a straight line. The first point of note is that there is no difference in quality between the information obtained under steady production conditions and that obtained during more erratic production. (Note also that it is energy that is plotted, not energy per tonne which would produce a graph curved in shape and used quite differently).

There are three features of the graph that should be examined in detail—the slope, the fixed losses and the scatter.

The Slope

This represents the energy actually involved in making the product. In the case of a drier or evaporator it is the heat needed to evaporate water. In the case of a furnace it would be the energy needed to heat up the product. For many processes, particularly heat processes, it is possible to calculate this energy and compare it to what is found.

The Fixed Losses

The point where the line cuts the energy axis corresponds to the energy used under conditions of zero production. A value has been obtained here by extending the line to zero production. There is bound to be some uncertainty, if only because of the scatter in the points. In some cases, by taking advantage of short production pauses, it is possible to make a measurement of these losses. If so, this can add considerably to our understanding of the information we collect.

The Scatter

Initially it is the scatter to which most thought needs to be directed. Our interest is not purely statistical or in producing better data. This information could be run through a computer to find the formula for the line and the so-called confidence limits, which quantify the scatter. The real point of interest is the fact that in the best week (week 19) energy use was a good ten per cent

FIGURE 2.1: REPRESENTATIVE PRODUCTION AND ENERGY CONSUMPTION DATA

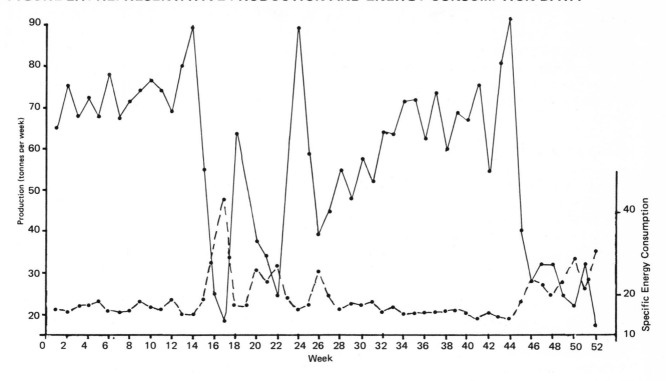

FIGURE 2.2: ENERGY USE IN THE FIRST SIX MONTHS

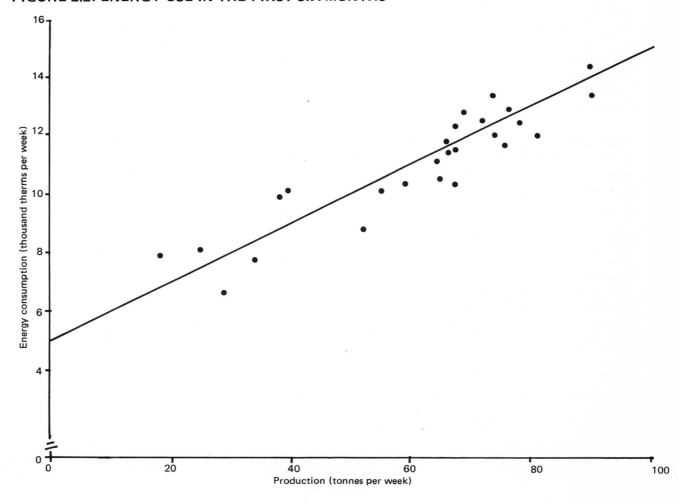

better than the average and in the worst weeks (17, 20, 9) it was five per cent worse than average. What is more important to know is whether the figures for week 19 are accurate: was it a fluke or could this be achieved consistently?

Scatter can be due to errors in the way production or energy are measured. If the product is not actually weighed but is counted as, say, numbers of bars of metal, part of the scatter will be due to variations in bar weight. Scatter can also be caused by poor control. The removal of product could be taking place at too low or too high a temperature, or it might be over dried.

There are several variations in the form of this graph that can occur as a result of special circumstances. These are shown in Figure 2.3. Type A is the most common. Type B has no fixed losses. It is fairly uncommon but has the advantage that energy per unit of output does not depend on production rate, so that monitoring is fairly straightforward. Type C has a very high fixed loss compared to the slope and usually indicates that something is wrong with the process and the fixed losses should be investigated. Type D occurs where the rate of production does not vary. There may be fixed losses that can be reduced but these cannot be quantified without specific investigation.

Type E is the same as Type A except that there is a very high scatter. This could be because the information on production and the information on energy use do not match, or it could indicate poor control of the process. Either way it requires urgent investigation. Type F is also like Type A but the line is curved. This is most common when looking at energy use of an entire site.

Measures applied to plant can alter the slope, the fixed losses or the scatter or any combination of these. Insulation usually only affects the fixed losses; heat recovery will affect the slope and the fixed losses; improved process control will usually affect the slope and the scatter.

Graphs of this sort should be plotted out for any unit of plant for which production figures and energy use are available. This includes boiler plant where fuel used should be compared to steam produced on a regular basis. They can be produced for overall site consumption as well but are less easy to interpret.

Relating Energy Use to the Weather

The main parameter of interest in monitoring the energy use of a building is the weather. Weather can affect the building in a number of ways: the sun beaming through the window; heat losses by conduction through the walls, windows and roof; and air losses through open doors, windows and gaps in the building fabric. It turns out that the most important are those which are directly related to the difference between the inside temperature of the buildings and the external air temperature. Clearly, it is only necessary to supply heat to the building when it is colder outside than the comfort temperature, and what is needed is a measure which will take this into account. Such a measure is the **degree day**.

On any one day the temperature is not constant but varies by several degrees. The degree day is a measure of by how much and for how long the temperature is less than a given base temperature on a given day. If this is added up for all the days in a given period we have something by which to monitor energy use in the building. To be absolutely accurate we ought to measure the outside temperature continuously but in practice that turns out to be much too involved (although microprocessor-based temperature recorders have been developed recently for the purpose). Instead, degree days are calculated from a formula based on the daily maximum and minimum temperatures. A set of degree day figures is published each month in *Energy Management* for each of 17 weather recording centres around the country and on ENTEL, the Energy Efficiency Office information service on Prestel. There is also a Fuel Efficiency Booklet available on degree days.

Using degree days is exactly analogous to the method we have described for monitoring energy use against production to establish a baseline. This should appear as in Figure 2.4 with a fixed energy use which will comprise that part of the energy use not directly related to maintaining the temperature of the building (canteen, hot water for washing, etc) and a variable degree day related demand. The scatter can often be greater on these degree day graphs, especially when it is possible

FIGURE 2.3: DIFFERENT FORMS OF RELATIONSHIP BETWEEN ENERGY
AND PRODUCTION

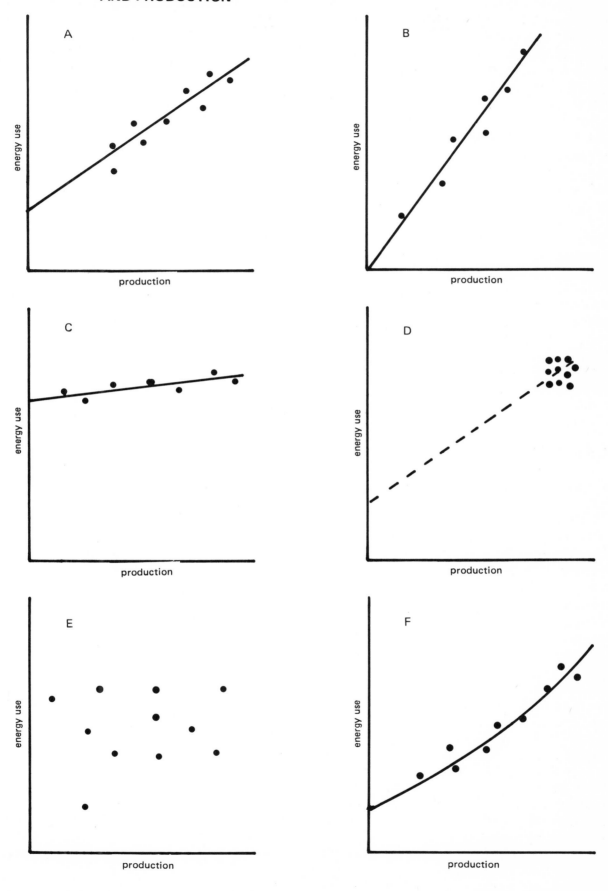

FIGURE 2.4: ENERGY USE RELATED TO DEGREE DAYS

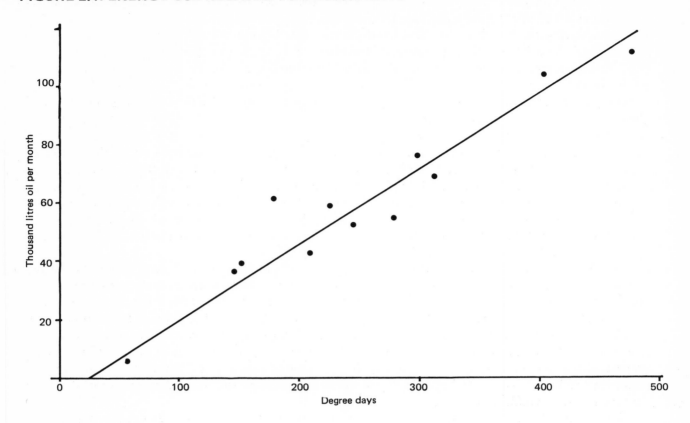

TABLE 2.5: EVALUATING CUSUM

1 Week	2 Production (tonnes)	3 Energy Used (therms)	4 Predicted Energy (therms)	5 Difference Col 4—Col 3	6 Cumulative Difference (CUSUM)
1	65.00	10.5	11.5	+1.00	+1.00
2	75.5	11.6	12.55	-0.95	-0.05
3	67.5	11.5	11.75	-0.25	-0.20
4	72.00	12.5	12.2	+0.30	+0.10
5	67.5	12.3	11.75	+0.55	+0.65
6	78.00	12.4	12.8	-0.40	+0.25
7	67.5	10.3	11.75	+0.55	-1.20
8	71.5	11.4	12.15	-0.75	-1.95
9	74.00	13.4	12.4	+1.00	-0.95
10	76.5	12.9	12.5	+0.25	-0.70
11	74.00	12.0	12.4	-0.40	-1.10
12	69.00	12.8	11.9	+0.90	-0.20
13	81.00	12.0	13.1	-1.10	-1.30
14	90.00	13.4	14.0	-1.00	-1.90

for the occupants of the building to affect energy use by uncontrolled action—by opening windows, interfering with thermostats, using supplementary heating such as electric fires, etc.

Occasionally, the line will cut the degree day axis of the graph instead of energy consumption as shown here. This is because the heating system may be controlled so that it does not come on until the outside temperature drops some way below the temperature chosen as the degree day base temperature.

One aspect of degree days which gives rise to some puzzlement amongst users is the choice of base temperature. Normally a heated building in the UK would be regulated to $18.3^{\circ}C$ ($65^{\circ}F$). This is above the minimum temperature required by the Factories Act 1961 of $15.5^{\circ}C$ and the Shops and Offices and Railway Premises Act 1963 of $16^{\circ}C$ but below the maximum set by the 1980 Fuel and Electricity (Heating Control Amendment) order of $19^{\circ}C$. Published degree day tables are usually for a base temperature of $15.5^{\circ}C$. The reason for this is that not all the heat in a building comes from the heating plant—some heat is generated by people, lights, and electrical equipment such as refrigerators, kettles, photocopiers. Also, because of the thermal characteristics of a building, the need to supply heating can often be avoided until the outside temperature drops a little below the control temperature.

Established practice assumes that the best way to handle this is simply to calculate heating degree days to a slightly lower base temperature. This new base temperature is then considered to be **occupancy corrected** and the $15.5^{\circ}C$ value commonly used is $18.3^{\circ}C$ with a correction by $2.8^{\circ}C$.

Hospitals commonly work with a base temperature of $18.5^{\circ}C$ on the grounds that temperatures need to be kept higher than in an office to be comfortable to patients who are less active. Moreover, not a great deal of heat is generated in the hospital. In general, the selection of the base temperature is something that need only be considered in the light of experience. The precise choice of base temperature is not critical and should not be allowed to occupy too much discussion early on in an energy saving programme.

The published degree day data currently available are only for heating degree days to base $15.5^{\circ}C$ and $18.5^{\circ}C$ for calendar months for 17 weather centres around the country. Some users of degree days find they need values calculated differently. There are three main circumstances under which this might happen.

Geographical Location: Local factors could give reason to suppose that the pattern for the locations of interest is different from that of the nearest of the 17 published stations.

Time Periods: The published tables are for full calendar months. Some buildings are not continously occupied (e.g. schools), or energy data may only be available for some different period (e.g. accounting periods).

Cooling Degree Days: In some buildings the aim may be to maintain the temperature below a maximum temperature rather than above a minimum temperature for part or all of the year, e.g. for air conditioned buildings or cold stores. For these, use of cooling rather than heating degree days is necessary.

Time Trends

Having established the relationship between production and energy use, or degree days and energy use, for a given period as a baseline, it is then possible to compare subsequent energy use to that baseline. The most sensitive technique is known as CUSUM. CUSUM is an acronym for *cumulative sum* and is a measure of the progressive deviation from the previous consumption pattern.

CUSUM is simple to calculate and has the advantage that it can often be incorporated into the existing reporting system as shown in Table 2.5. This shows a log-sheet such as might already be in use. The first three columns will almost certainly be recorded already.

There are three added columns. Column 4 is a predicted energy consumption figure determined from the baseline for the particular production level either by reference to the graph or, better still, by using a formula calculated from the graph. Column 5 is the difference between the actual energy consumption and that predicted in column 4. In column 6 this difference is added up week by week. This is a simple enough calculation to do by hand. It does not need a computer but certain spreadsheet programmes can make it quicker.

The entries in column 6 can then be plotted on a graph as shown in Figure 2.6 to give a visual impression of the savings made. More importantly, it enables separate measures for improving energy efficiency to be evaluated. For each measure implemented there will be a straight section of the CUSUM graph which begins when one measure was introduced and extends to the point where the next measure is introduced. Plotted in this way the total contribution to savings from each measure can be assessed at any time.

CUSUM can also be applied to energy consumption related to degree days. Column 2 would then contain the degree day value for the period, either obtained from published tables or measured locally. Otherwise the columns would be the same.

Sometimes energy consumption could be determined by both production and degree days when it might be necessary to derive a prediction formula of the form:

Energy = A + B x production + C x degree days

which would require the use of a computer to establish this formula from past consumption records. In this case the use of a computer programme, such as a spreadsheet programme like 'Supercalc' is usually necessary.

FIGURE 2.6: EVALUATION OF SAVINGS USING CUSUM

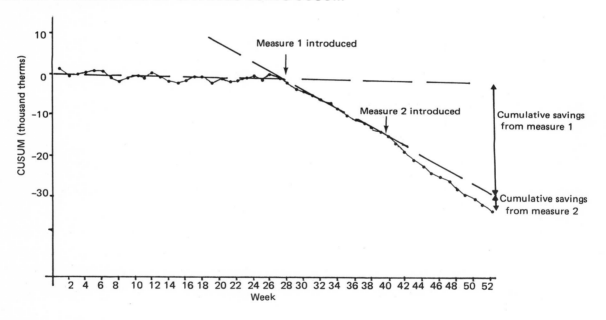

Monitoring and Target Setting Scheme

The Energy Efficiency Office is assisting a number of trade and research associations to develop energy monitoring and target setting systems in their respective industries. Pilot studies have been completed in the textiles and paper industries. Their success is leading to substantial savings, and has resulted in the British Paper and Board Industry Federation and the British Textile Employers' Association recommending implementation throughout their respective sectors. Projects are also underway in the bricks, pottery, steel, steel castings, non-ferrous metals, iron forging, timber, chemicals, food and drink sectors. Other sectors will be covered in due course.

3

MOTIVATING STAFF

Peter Martin

Any energy manager, but especially one newly appointed, can be forgiven a feeling of loneliness when he contemplates the business pressures within his organisation, and realises that it is his task to change the attitude and performance of employees, many of whom may be largely indifferent to his major pre-occupation: the cost effective use of energy. How is he to set about his task of motivating both workforce and management? The answer in principle is easy, the practice of course more complex. But one thing is certain, unless the energy manager is working from a sound set of principles, he has little chance of success.

The purpose of this chapter is to suggest the principles which should guide his actions. It considers whether or not people have a role to play, describes how he should consider the total environment or situation in which he operates, and considers the why, what's and how's that will be important to him as an energy manager.

From the outset it is necessary to discard the messianic approach. This is unlikely to work for long. People will undoubtedly respond with fervour, initially, to some ringing declaration of the importance of energy conservation. For a time they will turn out the lights in rooms that are not being used, and report dripping taps. Energy consumption will fall for a while, but something else will soon attract people's attention, and after a month or two it will begin to drift upwards once again. That has been shown time and time again.

The Need for Motivation

This phenomenon might suggest that people do not matter and that what is required is to install electronic systems and controls. That done, energy should take care of itself. Such a view would be wrong, but for reasons that are not immediately obvious.

Sophisticated systems clearly have a part to play. But they have one fault: they are incapable of thought. They do not have the ability to acquire insight and understanding into the why's and wherefore's of any particular operation. Only people can do that, and only people—and frequently only those involved with the process day in and day out—have the capacity to think how things might be changed or improved; and about how some fundamental rethink about purpose or design could alter total energy requirements. Put it another way. Control systems clearly are incapable of considering how the total operation might be changed and improved.

Basic Parameters

What about the attitude of people within the organisation? For example, if energy is a very small part of operating costs, then it is going to be harder to create interest in conservation than if it is a significant part. Or, if the firm is starved of investment funds, then ideas of investing in energy saving equipment can be placed on a back burner.

Information

Questions have to be asked about the availability and dissemination of managerial information. If management at operating level does not know the scale of their energy costs, it is going to be extremely difficult to motivate them. They need information or they have no idea how they are doing, and if they do not know, why bother? In other words, if the organisation is weak in financial systems then energy management is likely to prove an uphill struggle.

Investment Policy

Investment policy has to be established. To what use is the firm going to put its available capital? If it is engaged in a new building programme then maybe that is a key area for the energy manager to review. He may decide he should concentrate on the planning stage and ensure these buildings are as energy efficient as possible. If the buildings are old but new plant is being installed perhaps he should concentrate on its implications for energy consumption, bearing in mind the effect it will have on the patterns of production.

Management Style

He needs to consider also the prevalent style of management. Is it autocratic and largely indifferent to the view of the shopfloor? If so, he could probably find disaffected employees will have little incentive to contribute. Alternatively is the style open, with management and staff talking and listening to each other? If this is the case, then he may be sitting on a veritable gold-mine of potential help, and he needs to formulate his objectives carefully before running the risk of squandering valuable goodwill.

Key Questions for the Energy Manager

At this stage it becomes necessary to consider fundamental questions. If handled sensitively however, they can afford a useful focus for discussion with top management who are really the only people who can provide sufficient impetus for what the energy manager is trying to achieve. Thus armed with a realistic agreed policy, the energy manager is in a position to consider the why's, what's, have's and who's.

Firstly, why should energy management be considered important in his particular organisation? The answer may of course be obvious. For example, if energy is consumed intensively and claims a high proportion of operating costs, improvements will make an important contribution to profitability. Even if the potential saving is smaller it is still very likely to prove worthwhile.

The problem is that the time spent on energy management could be spent just as usefully on some other activity. It is advisable for the energy manager to spend some time identifying just how important controlling energy costs is likely to prove. For if energy saving appears as an important issue that will be a useful motivating factor when communicating with staff.

However, even if a lower degree of importance seems the case, there are some additional and, in some respects, unique points in favour of energy management. Firstly, investment in improving energy efficiency is risk free, and will show increasing returns as energy prices rise. That is, unlike an investment in sales or advertising, there is a very low risk of the project not displaying the due return. It is like investing in the pools with a guaranteed and increasing win every year.

Secondly, the subject of energy conservation provides an opportunity to look not only at energy but at the total process. So, for example, when Sainsbury's provided the question of how to make hot shrink wrap machines more energy efficient, they eventually came up with the solution that they would change to cold shrink wrap, and the savings were dramatic. The astute energy manager may wish to use energy as a pretext for viewing the total operation. This is a useful consideration when prodding colleagues and staff. They can take the credit for re-design; the energy manager can make the energy savings.

It is important to identify the key areas for effort. If the energy manager doesn't know them he will be unable to motivate anybody because nobody will know what they are supposed to achieve. And worse still, if the energy manager's 'what' does not seem credible, and other people think their ideas are better, they certainly will not be motivated his way. So it is important to know what is wanted, and to make sure it is the right 'what', or at least one that is extremely plausible.

FIGURE 3.1: DRAFT LETTER FOR OUTLINING SUPPORT FROM TOP MANAGEMENT

```
                                    Manor House
                                    Meldreth
                                    Royston
                                    Hertfordshire
                                    Telephone (0763) 60189 Telex 81383
```

```
To:        Members of the Management Executive

From:      The Managing Director



4th February 1985


Dear Fellow Director

Following our board discussion, I confirm that it is our
intention to achieve energy savings of 15 per cent in the
machine shop area over the next year, but without any loss
of output or quality.  We agreed that we should review
progress at the next board meeting in three months time.

Mr.Smith, Energy Manager, would be able to give technical
advice, and undertake much of the detailed work. I have
asked him to contact you in the next week.


Yours sincerely
```

Managing Director

The way to get to grips with what is wanted is, of course, to undertake an audit, something already covered in Energy Manager's Workbook Volume 1. That highlights the key areas for savings, and provides the basis for an acceptable effort.

The next matter to consider is whether the conservation effort should be 'sharp' or 'fuzzy'. In other words, should there be a concentration on a few sharply defined key areas for savings, or should there be a generalised overall attack, or perhaps a mixture of both? The answer depends on circumstances.

If the problem area is under the control of only a few people, and savings there would make a major impact, then clearly that is where the greatest effort should be concentrated. Scarce energy management resources should not be diluted in a generalised assault. If, however, we are talking about a consumption pattern to which everybody contributes—as in, for example, an office block— then the 'what' involves everybody. If the block is old and there is no money for investment, then maybe involving everybody is the only way, though the boilerman should clearly be earmarked for special attention. But the rule is simple. Concentrate your resources on key areas and do not waste time on going 'public' if that would be a distraction from the principal activity.

The Need for Top Management Support

The 'what' immediately leads on to the 'who'. Should the energy manager be motivating managers or staff? The prime rule is that the people to be converted are the management, and particularly the top management. There is a simple rule of thumb. If the people at the top are not behind energy management, then achieving results will be difficult.

The first task then is to obtain support from those at the top. It is perhaps a rather easier task than people realise. The way to proceed is as follows. The energy manager should be completely clear in his own mind what it is he wants to do. He must know his key action areas and targets and who amongst the management he wishes to persuade to take action. He then wants to put himself in the position of the Managing Director. If action is requested in important areas, he decides his objectives, and after discussion usually issues appropriate directions. This suggests that the way for the energy manager to proceed is to meet the Managing Director or an appropriate senior manager and draft for him a letter that he might want to sign. It could read along the lines of that in Figure 3.1. Note that the letter is in the form of an objective, and we shall come back to objectives later.

Once the first such letter has been written, the rest becomes easier. Provided the energy manager demonstrates professional competence, fellow managers will be only too eager to co-operate.

Tackling the Workforce

Once the decision has been taken to involve all, or a proportion of, the employees in an energy efficiency programme we must decide how best to go about this. If there is not already information to hand from which they can monitor any progress made, then the possibility of supplying such information should be investigated.

There are various means by which this might be achieved. For example, the use of posters might be considered most appropriate, or a conference could be seen as the way forward. This is the point at which it is necessary to look clearly at how the firm or the organisation works. The question 'do we have a committed workforce?' needs consideration. If the answer is no, the future is bleak, and energy management, like everything else in the organisation, is likely to prove an arduous task.

What are the signs? Sometimes the energy manager will already know the answer in his bones. If he is not sure, he might consider such matters as staff turnover and absenteeism, excessive overtime working, and so on. If there is no commitment, then his approach is going to have to be very much that of concentrating on key areas and relying on his own resources, and the resources of those managers who are prepared to help. Effort expended in an atmosphere of general indifference is frequently effort wasted. The only people who can start to turn the tide in this situation are those at the top and, except in special circumstances, energy management is unlikely to be their first priority.

Using Existing Channels

A high level of staff commitment should, however, be expected, which will enable progress to be made. The energy manager should use some of the channels that are already available. The word 'some' is used advisedly. Many organisations have elaborate consultation and participation procedures, but directors will admit privately that these are not effective, despite the elaborate arrangement of area, regional and central bodies and representatives of all sorts of groups and a regular programme of meetings. For example, one company—a household name but which must obviously remain anonymous—stated privately that in the middle of the most momentous events, including a major redundancy programme, their central consultative council confined itself to discussing the provision of bicycle sheds!

The lesson to be learned from this is that the energy manager needs to know the channels of communication that actually work. There is one for example that is being used increasingly by companies such as Jaguar and Vickers. This is the briefing group that 'cascades' down the organisation. The form varies. In one version, each director briefs his senior managers, they in their turn brief the middle managers, and so on down to the supervisors addressing the shop floor. In another, the department is addressed not by its manager (father), but by his manager (grandfather) in the 'father's' presence. In others the Managing Director makes a video, and the departmental manager or supervisor discusses the presentation after it has been shown.

The Value of Information

Throughout companies there is a trend to disclose far more information than before. In some companies information is now available that a few years ago would not have gone outside the board room.

Energy managers should latch on to these trends. They should arrange for information to be fed into briefing groups, and perhaps make themselves available to answer questions at some of the groups more important to their cause.

Quality Circles

The energy manager needs also to be aware of the development of quality circles. Admittedly some of these do not work particularly well. But in the best, a real effort is made to involve employees constructively. The theory is that there can be nobody who knows more about a job than the employee who does it day in and day out. His or her knowledge is invaluable, and if it can be harnessed, the company can only benefit. In these circles groups of employees meet and discuss ways in which they can improve the operations that concern them. If good circles exist, the energy manager should participate in them, feed in his energy problems and ask for advice. He will be pleasantly surprised by the answers he receives and the help he is given.

Benefits for the Employee

The energy manager should always be conscious of the question 'what is in it for me?', the question that is posed, albeit unconsciously, by all managers and staff. Sometimes the answer is easy. For example, a few firms have value-added schemes. A percentage of any cost saving goes into the employee's pocket. In others, there are employee share schemes, and in some companies it is reported that employee attitudes have improved significantly since the introduction of such schemes. Other companies have bonus schemes related to return on capital. The employee incentive is obvious. Some companies simply have such good management that the employees co-operate as a matter of course. The course that the energy manager should follow is to find out how employees are motivated and harness the organisation's structure to his cause.

It is hoped that it is now evident how difficult it is to describe how the energy manager should go about his task. He must be like the good soldier and use the weapons to hand. Sometimes he will be lucky and find a ready audience. On other occasions he may not be so well served. But he can console himself, sadly, with the thought that if the energy message is especially hard to put over, so probably will be every other message in the company and the organisation will certainly not be firing on four cylinders.

Setting Targets

Finally, in setting himself targets, or agreeing them with others, the energy manager must remember that every valid target has six qualities, and if any one of them is missing the target is likely to become inoperable. The six qualities are:

- —the targets must be measurable

- —be achievable

- —be time based

- —be capable of being monitored

- —maintain a given standard

- —there must be agreement between the manager and subordinate as to how efforts towards achieving the target are to be supervised.

What all this means is that a measurable target is one expressed in numbers, for example, reduction from 8,000 kW to 6,000 kW. Achievable means what it says. If the target is not capable of being realised, then it is not an effective target. Nobody will try to achieve it. In an ideal world the target should stretch somebody, but not to the point where attainment is impossible. The world is not ideal, and setting targets with that degree of precision is difficult, but the effort is worthwhile.

The time by which the result is to be achieved must be stated explicitly. If it is not, the commitment is open-ended and there can be no measurement of progress. It must be possible to see how much progress is being made. The result cannot be allowed to pop out at the end. If a target is being monitored, it becomes possible to take corrective action if things are going adrift.

'Maintaining a given standard' is frequently forgotten, but is important. For example, it is quite possible for a lighting engineer to reduce consumption from 5,000 kW to 3,000 kW, and thus apparently achieve his target. The result is not very satisfactory if the office then becomes so dim that people have difficulty in reading. The correct target should have included a statement like 'reduce consumption from 5,000 kW to 3,000 kW, but maintain an intensity of illumination of x lumens per square foot.'

There should be agreement on how the subordinate should be supervised. This means, for example, that the employee should report to his boss, say every week, or when some subtarget has been achieved, or a quarter of the way through, or when he encounters certain sorts of difficulties and so on. The purpose of this agreement is to make it easier for boss and subordinate to work together so that the boss does not feel he is breathing down the subordinate's neck, or the subordinate does not feel worried about when he should approach the boss. Experience has shown that this sort of 'contract' plays an important part in making targets an effective form of management.

This is where correct drafting of the Managing Director's letter (Figure 3.1) is important. All six qualities are there, and by agreement in this particular situation the colleague will just 'get on with the job' until the next quarterly board meeting to monitor progress.

4

UNDERSTANDING ELECTRICITY COSTS

David Yuill

The whole field of electricity tariffs and charges is beset by complicated detail which tends to obscure the relatively simple underlying principles. This chapter offers to the non-technical user of electricity an understanding of those principles so that opportunities for cost reduction and the more effective use of electricity may be identified. With an appreciation of the electricity supply industry's varying costs and charges, the structure and detail of electricity tariffs becomes less puzzling.

This chapter starts with an examination of these costs and proceeds to how they are reflected in electricity tariffs. On the way some of the important technical terms are explained, and the chapter finishes by looking at ways of getting better value for money from electricity.

The Cost of Electricity

The unit of electricity or kilowatt hour (kWh) is the amount of electrical energy consumed for instance by a 1 kW electric radiator in an hour or by a 100 W lamp in ten hours. The consumption of electricity supplied by an electricity board is measured in these kWh units so that the consumer can be charged accordingly.

Although fuels such as gas, oil or coal have prices per therm, litre or tonne which vary from time to time there is no difficulty in knowing at any given time how much these units of energy cost. The same is not always true of electricity. Overall costs per unit vary considerably, according to time and pattern of consumption and the details of the tariff under which the supply is provided. Many consumers have overall average unit costs in the range 4-6 pence, but some supplies work out below this range and others can be a great deal higher. Figure 4.1 is an extract from an actual monthly bill showing an average cost per unit of 21.7 pence. There were in fact special reasons for this unusually high average cost, and Figure 4.2 shows a more typical winter bill from the same electricity board. The average unit cost of this second bill is 4.5 pence per kWh.

Comparison with Other Fuel Costs

It is not always easy to make a direct cost comparison between electricity and other fuels used for heating. The wide range of average unit costs indicated above is one problem, but there are also difficulties in that some electric heating techniques are significantly more effective than established fuel heating methods. However, if certain assumptions are made, for instance a unit cost of 4.5 pence and an efficiency of 75 per cent for the conversion of fuel into useful heat, then comparative costs per useful therm can be calculated, such as those shown in Table 4.3.

Electricity is often available for 'off-peak' heating, e.g. between 0030 and 0730 hours at around 1.7 pence per kWh. This gives useful heat at 50 pence per therm, i.e. very near to gas at practical efficiencies of conversion in boilers etc.

In practice there are many applications of electricity for heating where costs can be attractive compared with other fuels. Whether electricity is used for heating or in its unassailable roles of lighting and motive power the costs can vary greatly, and the principles underlying these cost characteristics need to be understood.

FIGURE 4.1: A MONTHLY ELECTRICITY BILL SHOWING A LOW LOAD FACTOR

METER READINGS *		Consumption *	METER READINGS *			Consumption *
Present	Previous		Present	Previous		
000237	000237		558798	556299	R	2499
UNITS X	2.00	0	142423	141877	N	546
000587	000587					
UNITS X	2.00 N	0				

MAXIMUM DEMAND THIS MONTH	K.V.A. (Daytime) 123.0	K.V.A. (Night time)	ANNUAL MAXIMUM DEMAND	K.V.A. 150.0	Month Recorded SEPT

DESCRIPTION OF CHARGE	No. OF UNITS OR K.V.A.	RATE	AMOUNT EXCLUSIVE OF TAX	VAT REG. No. 238 5679 21	
				Tax	% Rate
ANNUAL KVA 1	100.0	£9.960/12	83.00		00.00
ANNUAL KVA 2	50.0	£8.580/12	35.75		00.00
MONTHLY KVA	123.0	£3,560	437.88		00.00
UNIT CHARGE	2499	3.597P	89.88		00.00
NIGHT UNITS	·546	1.850P	10.10		00.00
FUEL CLAUSE	3045	0.16500P	5.02		00.00
		TOTAL	661.63	0.00	

NEXT NORMAL METER READING	M.D. THIS MONTH (KW)	COST OF FUEL PER TONNE	£		
DATE:- 27 JAN	121.5	48.75		661.63	19 JAN 1984
* above 'E' = ESTIMATED. 'R' = METER REMOVED 'N' = NIGHT UNITS				TOTAL DUE	LATEST DATE FOR PAYMENT

FIGURE 4.2: A MONTHLY ELECTRICITY BILL SHOWING A HIGH LOAD FACTOR

METER READINGS *		Consumption *	METER READINGS *			Consumption *
Present	Previous		Present	Previous		
88378	85204		LESS NIGHT UNITS			
UNITS X	10.00	31740	00487	00151		
			UNITS X	10.00	N	3360

MAXIMUM DEMAND THIS MONTH	K.V.A. (Daytime) 86.0	K.V.A. (Night time)	ANNUAL MAXIMUM DEMAND	K.V.A. 92.0	Month Recorded DEC

DESCRIPTION OF CHARGE	No. OF UNITS OR K.V.A.	RATE	AMOUNT EXCLUSIVE OF TAX	VAT REG. No. 238 5679 21	
				Tax	% Rate
ANNUAL KVA 1	92.0	£7.150/12	54.81		00.00
MONTHLY KVA	86.0	£3.420	294.12		00.00
UNIT CHARGE	28380	3.442P	976.83		00.00
NIGHT UNITS	3360	1.540P	51.74		00.00
FUEL CLAUSE	31740	0.12600P	39.99		00.00
FIXED CHARGE	100		5.10		00.00
		TOTAL	1422.59	0.00	

NEXT NORMAL METER READING	M.D. THIS MONTH (KW)	COST OF FUEL PER TONNE	£		
DATE:- 25 FEB	86.0	48.00		1422.59	14 FEB 1983
* above 'E' = ESTIMATED. 'R' = METER REMOVED 'N' = NIGHT UNITS				TOTAL DUE	LATEST DATE FOR PAYMENT

TABLE 4.3: COMPARATIVE FUEL COSTS PER USEFUL THERM

	Price	*Pence per therm supplied**	*Pence per useful therm*
Electricity	4.5 pence per kWh	132	132
Gas	37 pence per therm	37	49
Gas oil	22 pence per litre	61	82

*29.3 kWh are equivalent to 1 therm

Costs of Generation and Distribution

To a considerable extent the electricity supply industry tries to apportion the charges it makes to consumers so that these fairly reflect the costs of generation and distribution of electricity to each user. Looking at the operation from this point of view should help to explain the main features of tariff structures.

By making some assumptions and taking a simplified view of the coal-fired steam power generation process the fuel cost of producing a kWh unit of electricity can be roughly estimated.

Assuming that:

overall efficiency of generation = 30 per cent
cost of coal   = 17 pence per therm

Since one therm is equivalent to 29.3 kWh, the fuel cost is:

$$\frac{17 \text{ pence} \times 100}{29.3 \times 30}$$

$$= 1.9 \text{ pence per kWh}$$

This is the sort of price charged for off-peak or night units. The night units of the bill in Figure 4.1 were charged at 2.015 pence including fuel clause adjustment.

Of course there are many costs in addition to the cost of fuel, and many of these arise from the fact that the level of demand for electricity varies a great deal. Generating plant and distribution facilities have to be provided to meet the highest levels of demand, but much of this capacity is not working, or earning money, most of the time. The fact that loads vary, often with little notice also increases the costs since generation in these circumstances is less efficient. Because the most efficient generating plant is run for as many hours as possible, i.e. it carries the base load, it follows that the average efficiency of generation is also lower at times of peak demand.

It is not necessary to estimate these costs of stand-by facilities and reduced efficiency, but simply to accept that they exist and are very significant.

Load Factor

The load factor of a particular supply may be regarded as a measure of the extent to which the available supply has been used; alternatively it may be taken as the average level of demand compared with the highest level of demand in a period.

This can be illustrated by looking at a simplified domestic load over 24 hours, see Figure 4.4.

FIGURE 4.4: DOMESTIC LOAD FACTOR OVER A 24 HOUR PERIOD

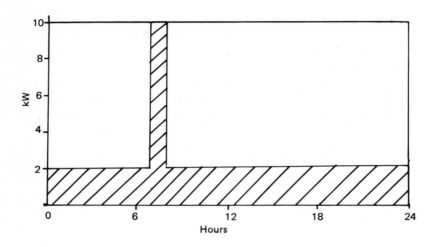

This 'demand profile' shows a load which is constant at 2 kW throughout the 24 hour period, except between 0700 and 0800 hours when it rises to 10 kW.

The 24 hour load factor   = $\dfrac{\text{total kWh units x 100}}{\text{maximum demand kW x hours in period}}$ per cent

= $\dfrac{56 \times 100}{10 \times 24}$ per cent

= 23.3 per cent

From what has already been said it is clear that consumers who have large peaks of demand and low total unit consumption, i.e. a low load factor, may be considered responsible for a major share of stand-by and other costs related to 'peaky' demand.

The load factor for the month covered by the electricity bill in Figure 4.1 is:

$\dfrac{\text{total kWh units x 100}}{\text{MD in kW x hours in period}}$ per cent

= $\dfrac{3045 \times 100}{121.5 \times 31 \times 24}$ per cent

= 3.4 per cent

This very low monthly load factor is consistent with the unusually high overall average unit cost of 21.7 pence.

The monthly load factor in the bill shown in Figure 4.2 is 50 per cent and the average unit cost is 4.5 pence. The bills are not strictly comparable in all respects, and are a year apart in time, but the inverse relationship of average unit cost and load factor is evident.

It must not be assumed that a high load factor necessarily implies efficient use of electricity, since a 100 per cent load factor would be achieved by switching everything on and leaving it on all the time. It does, however, produce a relatively low average unit cost on most non-domestic tariffs.

Maximum Demand

Since peaks of demand impose unwelcome costs on the electricity supply industry, most tariffs for medium and large commercial and industrial supplies include a charge related to maximum demand as well as for the units consumed.

For this purpose in the UK maximum demand is defined as the highest average rate of consumption in any half hour period, or twice the number of units metered in any half hour period. This means that a single very brief surge of power consumption does not itself establish liability for charges at that instantaneous level of demand.

Maximum demand is measured by some electricity boards in kW, and by others in kVA. The distinction between these two quantities and their relationship is discussed under the heading Wattless Current, later in this chapter.

Demand Metering

Electricity meter installations vary a great deal in detail, but where charges are based on maximum demand tariffs there are certain common functions. They all need to measure the unit consumption and the level of maximum demand.

FIGURE 4.5: KW DEMAND METER

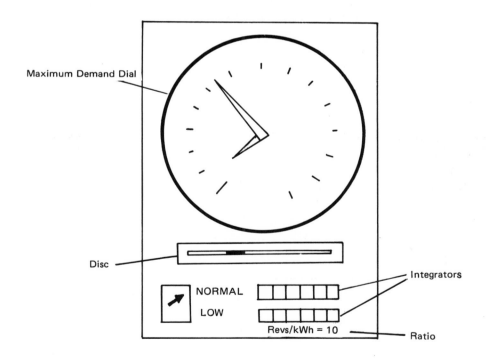

Typical features of a kW demand meter are shown in Figure 4.5 and include:

A kWh integrator shows the number of kWh units consumed. The difference between successive monthly readings is the monthly consumption for billing purposes. Day units (normal) and night units (low or off-peak) may be shown separately where the consumer has so requested. Although provided primarily for charging purposes, these integrators can be used by the consumer for monitoring consumption or for other energy audit work.

A circular horizontal disc may be seen revolving, and the ratio revs per kWh will be shown on the meter. A spot on the edge of the disc enables the number of revolutions in a minute to be counted. Multiplied by 60 and divided by the revs per kWh factor, this gives the instantaneous kW level of consumption. Such a spot check is often useful in energy audit work.

The maximum demand dial is marked in kilowatts. There is a large pointer and a much shorter one. The short pointer moves round the scale every half hour, adding up the units consumed. It starts from zero and returns sharply at the end of each half hour period. The point on the scale reached by this small pointer by the end of the half hour is twice the number of units metered. The long pointer is pushed round the scale by the small pointer, and remains at the maximum recorded level until reset to zero by the board's meter reader.

A meter factor such as 2, 6, 10 etc is often incorporated, by which all readings are to be multiplied to give actual kW and kWh quantities.

Wattless Current

For many uses of electricity where the load is, or behaves like, a resistance there is a simple relationship:

$$\text{power} = \text{voltage} \times \text{current}$$
$$\text{or} \quad \text{kW} = \frac{\text{volts} \times \text{amps}}{1,000}$$

This relationship applies to loads such as space heating, water heating and most lighting.

However, there are loads where this is not the case, notably in electric motors, arc welding, induction heating. These are all applications involving alternating magnetic fields, and current flows in these circuits in addition to the current which produces heat or motive power. This may be called wattless current or reactive current. This magnetising function does not itself consume any energy. The wattless current returns to the board's supply system and is not registered on the kWh meter.

The presence of wattless current in the generating and distribution system has the effect of increasing the total current flowing, so the current carrying capacity of the system has to be greater, i.e. requiring heavier cables, transformers, switchgear etc. Since all conductors have some resistance, there is also a loss of energy in distribution attributable to this extra wattless component of the total current. Electricity boards therefore generally want to identify those consumers who are responsible for significant levels of wattless current and charge accordingly, at least where the proportion of wattless current is unacceptably high.

Electrical engineers use a number of analytical techniques, for instance vector diagrams, to explain the relationship of power, reactive current etc, but for those who do not wish to master a great amount of theoretical detail a simpler analogy may be helpful.

Figure 4.6 shows a boat with a speed of four knots crossing a river, aiming at right angles to the bank. It is carried off course by the river current of three knots, and the resultant motion is at five knots as shown. A right-angled triangle has been used here to combine two currents and produce a resultant or total current.

FIGURE 4.6: BOAT ANALOGY AND POWER TRIANGLE SHOWING THE INFLUENCE OF WATTLESS CURRENT ON THE POWER FACTOR

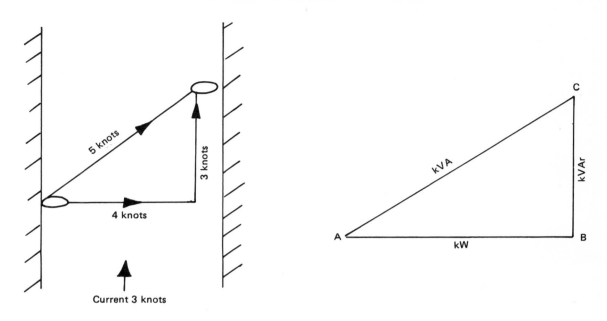

The combination of electric current which produces power and the associated wattless component of current in the same circuit may be represented by a similar right-angled triangle—see Figure 4.6. The side AB represents the current which produces power measured in kW or 1,000's of watts. BC is the reactive component measured here in kVAr, i.e. 1,000's of volts x reactive amps. The hypotenuse AC of the triangle represents the combined, resultant or total current in the circuit and is measured in kVA or 1,000's of volts x total amps.

It is necessary to understand this relationship of the two components kW and kVAr and the resultant kVA because about half the area electricity boards in the UK measure maximum demand in kW and the rest in kVA. This distinction is apparent when looking at tariff details or actual electricity bills.

Power Factor

The power factor in any circuit is the ratio $\dfrac{kW}{kVA}$

It is sometimes described as $\dfrac{\text{real power}}{\text{apparent power}}$

From looking at the various power triangles in Figure 4.7 it can be seen that power factor can never be greater than unity, and also that a high proportion of wattless current is the same thing as low power factor.

In those electricity board areas where the maximum demand is measured in kW the proportion of wattless current may be indicated on the bill by stating the power factor, or by comparing the kWh and reactive units recorded by meters. A meter installation for this type of tariff would include a kW demand meter similar to that in Figure 4.5 and a kVArh integrating meter.

On the other hand, in those areas where maximum demand is measured and charged for in kVA the minimum metering requirement is a kWh integrator plus a kVA demand meter.

FIGURE 4.7: POWER TRIANGLES FOR VARIOUS POWER FACTORS

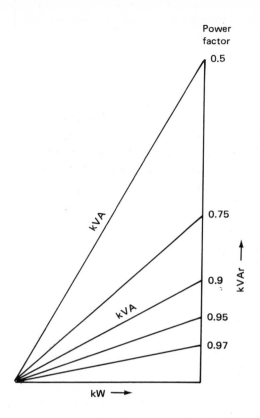

Low power factor means excessive wattless current, and in most board areas it leads to increased charges over and above the charge for the maximum level of power consumption. It is likely to occur where there is a high proportion of magnetising load, particularly where large electric motors run lightly loaded.

Low power factor or excessive wattless current can be corrected by installing power factor correction equipment at the point of supply or use. Electricity boards or specialist contractors will quote for the cost of such power factor correction and estimate the annual savings from reduced charges.

The optimum level of power factor after such correction depends on the tariff details and varies from site to site. Power factors of much less than 0.9 generally warrant investigation.

Supply Voltage

Consumers with demands of 500 kW or more may have an option of taking their supply at high voltage—6,600 V or 11,000 V—before transforming down to 415/240 V. In recognition of the advantages which this holds for the board, and of certain additional costs and losses incurred by the consumer under such an arrangement, the board's charges are generally less. Details vary from area to area, and thorough investigation and discussion with the board's staff are advisable.

Electricity Boards

There are twelve electricity board areas in England and Wales and two in Scotland. They each produce their own tariffs for the sale of electricity. These tariffs have similarities in principle, particularly because the English and Welsh boards all buy their electricity from the Central Electricity Generating Board under the same bulk supply tariff. There are, however, wide variations in the detail of metering, charging and terminology.

All boards have staff to advise consumers on choice of tariff and on the economical use and control of electricity, and this is a service of which full advantage should be taken. It is up to the consumer to make enquiries—boards are not generally required to proffer advice on such matters unless they are approached.

Types of Tariff

The many and varied tariffs are designed to recover the cost of fuel, installation, maintenance, servicing and other overhead costs of the electricity supply industry. In particular, many features are included to encourage the use and control of electricity in such a way that the industry's overall load is more even (i.e. it has a higher load factor) and therefore more efficient in terms of average unit cost. These features include cost penalties for peaks of demand, especially in winter, and price incentives to encourage use at night, weekends etc.

Non-domestic tariffs fall into three principal groups:

—general purpose or block tariffs
—maximum demand tariffs
—time of day tariffs

Each of these will now be discussed in turn.

General Purpose or Block Tariffs

These are for the smaller consumer, with maximum demand usually not exceeding about 50 kW. Maximum demand is not measured for charging purposes, and payment for units is in blocks of consumption during each period at reducing rates. The size of these blocks of units may be fixed in the tariff or is sometimes related to the connected load or potential maximum demand assessed in some other way, e.g. floor area.

General purpose tariffs for the smaller consumer usually offer some form of off-peak incentive which can be attractive for night use, and there are evening and weekend options which favour consumers such as the leisure or catering industries.

At the higher end of the range of consumption covered by general purpose tariffs, comparison should be made with the alternative maximum demand tariff. At some point this becomes cheaper, particularly if there is a reasonable load factor (i.e. maximum demand is well controlled).

Maximum Demand Tariffs

These cover the majority of industrial and commercial electricity consumers. Units are charged at fixed rates plus a fuel cost adjustment. There is usually an option to take night units, say between 0300 and 0730, at about half price or less, but some boards then charge more for the day units. In such cases separate day and night metering is only attractive if the night units exceed a certain proportion of the total.

A charge is made against the maximum demand recorded each month. This charge per unit of maximum demand is greater in the winter months, and in some areas nil from April to October; in others the summer months are charged at a lower rate.

Some boards offer 'prescribed hours' tariffs under which monthly maximum demand is only recorded and charged at the board's times of overall high demand, e.g. working daytime hours in winter. Such options may be attractive to leisure or cold storage operations, for example.

Maximum demand is either metered and charged in units of kVA or it is metered in kW and some additional charge is then made if the power factor is below a specified level—generally about 0.9— or if the kVArh units exceed a certain proportion, say 40 per cent, of the kWh units for the period.

A third main category of charge is related in some way to the likely maximum overall demand of a site. This monthly charge is specified at a certain amount per kVA. The kVA level is either fixed by agreement—often when the supply is installed—or it may be the highest demand recorded in the past twelve months (as in Figures 4.1 and 4.2). Availability charge or service charge are terms sometimes applied to this category.

Time of Day Tariffs

Recently, a few area boards have introduced optional alternatives to the monthly maximum demand tariffs. Instead of having mainly fixed unit rates plus significant maximum demand charges, these tariffs have no maximum demand charge but the unit price varies according to time of day and season. In one area these unit prices range from about two pence at night to as much as 30 pence for the two hours 1630 to 1830 on winter weekdays.

The increasing introduction of such time of day tariffs is restricted at present by the unavailability of suitable metering, but it is certain to become more widespread as part of the electricity supply industry's strategy to obtain a higher load factor.

Specially Negotiated Contracts

Special negotiation is theoretically possible if the range of standard tariffs can be shown to be inappropriate for a particular pattern of consumption. In practice, the opportunities for such negotiation are limited to loads which fit, or can be made to fit, into the times of low demand for the area or system as a whole. Large consumers who are prepared to cut their daytime demand at short notice may well be able to arrange special rates. Boards will sometimes offer modifications of their standard tariffs to secure particularly attractive loads for electricity rather than other fuels.

Choice of Tariff

For the great majority of consumers choice is limited to the published tariffs and options. Not all of these are included in the standard tariff leaflets and brochures which are issued from time to time. The possible existence of such options should be considered when seeking advice from an area board on the most suitable tariff for a known load pattern.

Boards can only advise on the basis of past consumption, and many have a computerised tariff comparison service based on the consumer's latest twelve months' billed consumption. If there is a possibility of modifying the demand pattern, it could alter the tariff recommendation. Such matters can be discussed with the supplying board free of charge.

The selection of the best tariff terms, and the reduction of consumption or chargeable maximum demand depend so greatly on the detail of the load pattern that a period of continuous load recording is an invaluable starting point. Scrutiny of such a demand profile record will often disclose opportunities which would otherwise remain unnoticed.

Load Management

Having chosen the best tariff conditions, electricity costs may often be further reduced both by avoiding superfluous consumption and by scheduling necessary consumption so that maximum demand charges are minimised and full advantage is taken of any off-peak rates.

Quite simple planning and programming can sometimes eliminate the coincidence of major loads at critical times. There are also many systems available which can anticipate periods of potential maximum demand and avoid these by shedding preselected parts of the load. A combination of programming and load shedding can secure major cost savings in suitable cases.

ENERGY CONSUMPTION IN BUILDINGS

Ken Spiers

Virtually half of all the energy used in the UK is for heating and lighting buildings. Consequently, designers and all those concerned with the operation of buildings need to understand the importance of energy costs, and must be able to plan and manage this resource efficiently. More than half of the total energy used in buildings goes into housing: the remainder being required for warming factories, garages, stations, airports, shops, hospitals, schools, offices, etc.

It is therefore in the field of space heating—and cooling—that attention can usefully be directed. There is no doubt that new buildings can be designed to use substantially less energy and that existing buildings can be significantly improved in this respect.

As the cost of energy increases there will be a progressive adjustment of resources toward improving features of the thermal design of buldings, e.g. improved insulation, electronic control systems, and a wide range of other measures giving increased emphasis to building design rather than the use of plant and energy to control the environment.

The energy consumption of a building is determined largely by the following factors:

—climate: temperature, solar radiation and wind

—volume of the building

—shape and orientation of the building

—area and position of the glazing

—insulation and thermal capacity (condensation)

—ventilation (infiltration)

—heat gains

—pattern of use

—comfort level

—efficiency of heat utilisation

—efficiency of heat generation.

The relative significance of the various ways in which building design can affect heat loss or gain will vary from building to building and even in the same building one decision may alter the significance of other factors, e.g. a decision to lower internal temperatures reduces the economic effectiveness of insulation. It is, however, very important for designers and building managers to be aware of the factors involved and of the approximate relative significance in typical cases.

Climate

There are few places in the world where human life can exist without some form of protection from the elements: the need for better environmental conditions is one of the major reasons for

FIGURE 5.1: FACTORS AFFECTING ENERGY CONSUMPTION

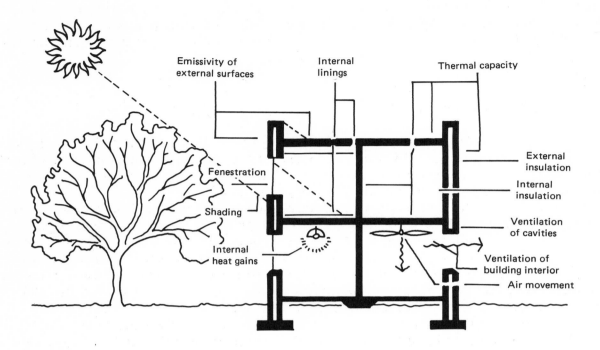

the existence of buildings. At its lowest level a building merely provides a degree of protection from rain and wind but many traditional building types, in extremely varied climatic conditions, have developed highly sophisticated thermal solutions, giving marked improvements in environmental comfort with very economical use of resources.

The courtyard houses of North Africa, the elevated wooden structures of many hot, humid climates, and the traditional English cottage have overcome very different environmental problems. They demonstrate with dramatic effectiveness how building form can control thermal performance. A Mediterranean courtyard house meets the problem of how to remain cool in a hot, dry climate with clear skies. The courtyards and their surrounding buildings radiate to the cold night sky and, during the night, a pool of cool air is built up in the courtyards and in the ground floor rooms. During the day the sun shines but the reservoir of heavy cool air remains for a considerable time. The walls of the building are of substantial thickness so that penetration of heat from the sun's rays on the walls takes some length of time. Ideally the heat will reach the interior during the night, when conditions are cool and the heat entering can be dissipated by ventilation. The walls themselves are painted white so that the minimum amount of solar heat is absorbed. In addition, in hot dry climates advantage can be taken of evaporative cooling: evaporation of water takes up significant amounts of heat and gives reduced air temperatures. The fountains in the courtyard therefore have a practical value in reducing air temperatures as well as a psychological attraction.

A very different problem exists in hot humid climates. There is no period of clear night sky to enable radiation to take place. The only available method for improving thermal comfort is increasing the velocity of air impinging upon occupants, both for its direct cooling effect and for the cooling which is still possible by more rapid evaporation of perspiration.

In higher latitudes overheating is no longer the main issue. During the winter heat input is needed and the problem becomes one of conserving the heat and giving as much habitable space as possible. Until very recently, timber, peat and dung were the only fuels available and their scarcity, together with the considerable amount of labour required for collection and delivery, meant that these fuels were expensive and could only be used sparingly. A typical traditional English cottage, with very small windows, very highly insulated roof, small volume owing to low ceilings, and a centrally sited fireplace and flue, exemplifies this.

Volume of the Building

The variation of heat loss with increase in volume is shown in Figure 5.2. It will be observed from this illustration that the increase in heat requirement is almost directly proportional to the increase in volume and that economical design to reduce space in the building yields major dividends in energy saving.

Shape and Orientation of the Building

The variation in heat loss resulting from diverse shapes of building is shown in Figure 5.2. Moving from the economical square plan shape (a) to the very different 1:3 aspect ratio with offset (e) results in an increase in heat loss of 20 per cent. Therefore a variation in shape can be a major consideration in designing a building with low energy consumption.

More extreme variations of shape, such as might result from transforming an aspect ratio of 1:1 into an aspect ratio of 1:10 must be rare and it is difficult to think of functions which can be adequately met by both plan forms without major variations in circulation area. This would mean in the case of the elongated form that the area of the external surface and the consequent heat loss would be substantially increased. With similar windows the ventilation rate would also be increased for most buildings. In addition, if the volume of the building were increased by additional circulation, the heat loss would be proportionately increased.

Extreme forms of these aspect ratios are often found in multi-storey buildings and here the heat losses are likely to be increased by the additional exposure of the high building. Energy costs are possibly further increased by the need for air conditioning because the height of the building and the lack of screening from trees and other buildings result in stack effects and solar gains necessitating air conditioning.

FIGURE 5.2: VARIATIONS IN HEAT REQUIREMENTS RELATIVE TO THE BUILDING PLAN

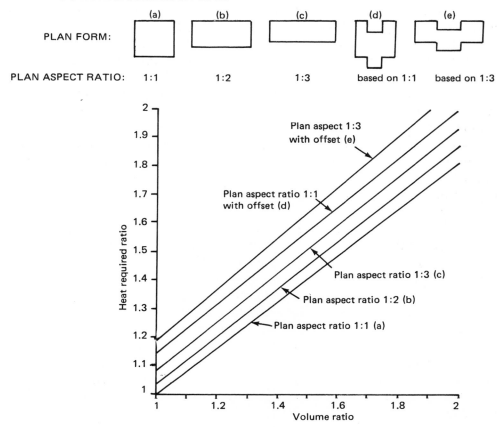

Area and Position of Glazing

Single glazing in a building is normally found to be the most sensitive part of the building fabric in terms of heat gain or heat loss. One moment the glass may be losing heat at the rate of 130 watts per square metre and the next allowing solar radiation into the building at a rate of 500 watts per square metre.

Solar heat can be utilised by manipulating the form, fabric and penetration of the building. Window glass is transparent to the rays of the sun which can pass through it and warm up the interior of the building. The glass is opaque to the radiation coming from the surfaces inside the building, but loses heat as a result of normal convection. If the radiation gain can exceed the convection loss then windows will contribute heat to the interior of the building.

The orientation of glazing is a particularly important consideration when designing a new building because it is possible to reduce the heat loss from the glazing, particularly by reducing the area of glass on the north face, and also to minimise the solar gains in order to prevent overheating during the summer period. The use of double glazing and solar control glass in modern building design allows a reasonable view out of the building, good comfort conditions and a reduction in installed heating and cooling plant capacity.

It would clearly be possible to adapt the design of buildings beyond the present conventional limits in order to take advantage of the full potential of solar gain. Lightweight finishes which would rapidly convert the solar rays to convected heat should be avoided. Massive elements of construction should be placed where the sun's rays can warm them. Orientation and window size should be carefully considered, together with screening and shuttering of windows to give reduced heat loss at night, and to control extremes of summer overheating.

FIGURE 5.3: EFFECT OF ORIENTATION ON HEAT LOSS

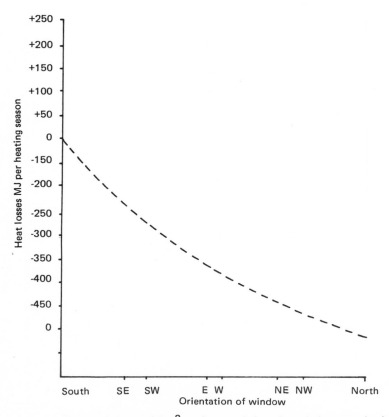

Net heat balance of 1 m² unobstructed clear glass window, curtained at night with 17.5°C mean internal temperature, for various orientations.

Figure 5.3 summarises the net heat losses for various orientations of one square metre of unobstructed window. It is apparent from this that efforts should be made in design to achieve the best possible orientations for windows.

A south facing window shows a gain during the heating season. It does not follow, however, that south facing windows should be increased in size for this reason. There are substantial periods without sunshine and additional plant would be required to maintain comfort during these periods. Larger windows may also contribute to overheating in summer.

Insulation and Thermal Capacity

In the days when a single open fire sufficed for home heating and it was commonplace to have most of the house unheated often with draughts and open flues, insulation would not have served a useful purpose. However, with the development of heating, higher levels of insulation became important, not so much to save energy as to provide comfort by ensuring an adequate surface temperature for walls and ceilings. The present boom in the sale of double glazing perhaps typifies the situation. It is very doubtful whether double glazing will save money, even at present fuel prices but it can give improved comfort. Insulation, in the present situation, does seem to present an obvious way of saving energy. In a badly insulated structure the effect of adding a layer of insulation will be dramatic. However, the effectiveness of adding further layers will progressively diminish.

The normal convention for assessing the thermal properties of buildings and for estimating plant size has been the concept of 'steady state' conditions. An estimate is made of the rate at which heat would be lost from the building, assuming the outside and inside temperatures remained constant over long periods. The figure resulting from this calculation, adjusted to take into account boiler efficiency, allowance for warming up etc, indicates the plant size required. Modifications of the calculation give an idea of seasonal heat requirements.

For many years buildings were characterised by small windows and heated mainly by means of low pressure hot water provided by solid fuel fired boilers. With this type of building and the slow response type of heating installation, the steady state method of analysis gave very acceptable results. For a substantial time, however, the thermal properties of buildings have tended to change towards lighter weight, which has a lower thermal capacity and consequently gives rise to much greater temperature fluctuations for a given heat input. New types of fuel, increased acceleration in the distribution system, improved emitters and much improved automatic control systems contribute to heating installations which are able to react very much more rapidly to changing circumstances.

New patterns of building usage exist, both in factories and offices, where intermittent occupancy of buildings calls for rapid warming up. It has become necessary to take a more sophisticated look at the way in which buildings behave thermally in order to relate the type of fabric to the type of installation and to relate both of these to the use required. It is important to take into account the insulation value of the materials of which the building is made, and also the capacity of these materials to absorb heat.

If the building is to be in fairly constant use, however, and a massive form of construction has been selected, there is probably little point in choosing an installation with sophisticated controls and rapid response, since these will serve little useful purpose.

It is not always necessary to select a complete lightweight construction in order to achieve rapid response to thermal conditions inside a building: an appropriate inner lining of low thermal capacity can transform a room from a slow to a quick response to heat input.

Table 5.4 shows the thermal capacity, (i.e. the amount of heat which can be retained), for walls of different materials having the same U-value. (The U-value is a measure of the rate at which heat is conducted through a medium for a given temperature difference across it e.g. a wall, door, or window, and is a property of the material of construction. The lower the U-value, the better the insulating properties).

TABLE 5.4: VARIATIONS IN THERMAL CAPACITY FOR CONSTANT U-VALUE

Material	Density (kg per m³)	Specific heat (J per kg per °C)	Volumetric specific heat (kJ per m³ per °C)	Conductivity 'k' (watts per m³ per °C)	Thickness for U of 1.0 watts per m² per °C (mm)	MJ required to raise this thickness for U of 1.0 by 20°C	Temperature rise resulting from application of 1 kW for 1 minute (°C)
Concrete	2,100	840	1,760	1.0	830	29.2	0.04
Brickwork	1,700	800	1,360	0.84	700	19.0	0.06
Timber	600	1,210	730	0.14	120	1.7	0.68
Lightweight concrete	1,000	1,000	1,000	0.3	250	5.0	0.24
Wood wool	500	1,000	500	0.1	83	0.83	1.4
Fibreboard	300	1,000	300	0.05	42	0.25	4.8
Expanded polystyrene	25	1,000	25	0.03	25	0.01	96.0

Note: A joule is the SI unit of work, energy and heat (equal to 0.239 calories). kJ is 10^3 joules, MJ is 10^6 joules, GJ is 10^9 joules.

The difference in thermal capacity is dramatically apparent. Upon this depends the speed of response of the buildings to fluctuations in heat input. The figures for warming up periods shown in the table demonstrate very effectively the differences in the time that would be taken to reach comfort conditions in rooms formed from the materials specified.

In buildings which are continuously heated there will be very little difference between the heat losses of different constructions having the same U-value. Some marginal differences may be discernible, since high capacity structures lessen the effect of control systems, but there will be little significant difference. However, in reality very few buildings are maintained at a fixed temperature day and night. Prisons and hospitals are the best known types which do have this pattern of heating.

Almost all other buildings have their heating either turned off or operating at reduced temperatures at night, and, in many commercial and office buildings, at the weekend also. At these times the buildings will cool down, losing heat not only through their fabric, but also by ventilation. They will then have to be warmed to comfortable temperatures to be ready for occupation.

If insulation is applied to the inner face of a building with a high thermal capacity, then the building will respond in the same manner as a lightweight building. This method of insulating a building is likely to create a problem with summer-time overheating and poor thermostatic control of space temperatures, and may even lead to interstitial condensation if a vapour barrier is omitted from the insulating zone. When insulation is applied to the outer face of a thermally capacious building, it is likely that the existing heating system and its controls will function satisfactorily, the summer-time temperatures will not reach an excessive level, and condensation will not occur.

Ventilation (Infiltration)

Ventilation is a major cause of winter heat losses and many of the public pronouncements on energy have called for reduced levels of ventilation. Figure 5.5 shows how, for a typical commercial building, heat loss varies with ventilation rate. Much could be done to decrease heat losses by reducing excessive ventilation. There is, however, a considerable risk of creating condensation problems by the restriction of ventilation. In building design there have been too many fundamental errors made by concentrating attention upon only one aspect, and making changes without fully considering their consequences. Before any attempt is made to achieve substantial reductions in ventilation rates, careful predictions and practical tests must be made.

In many buildings the air infiltration rate through the cracks in doors, windows, and building fabric is greater than the ventilation rate required even when doors and windows are kept closed. Under these circumstances major energy savings can be achieved by weatherstripping the various holes and cracks in the building fabric, and this will also result in an improvement in comfort conditions.

Internal Heat Gains

Heat gains are normally produced within buildings by lighting, bodies, solar radiation, process plant etc. Provided that the thermostatic temperature controls are located in the vicinity of the heat gains, it is possible to reduce the total applied heat by the value of the heat gains. When a building heat loss has been reduced by means of insulation or weatherstripping, then the heat gains provide a greater proportion of the heat requirements, and attention must be paid to type of temperature control installed, otherwise overheating and discomfort may result.

If a building is provided with an external temperature compensator type of control, any heat gains produced within the building will be in addition to the heat added by the heating system, and overheating is an inevitable outcome.

FIGURE 5.5: PROPORTION OF HEAT LOST THROUGH VENTILATION IN A TYPICAL OFFICE BUILDING

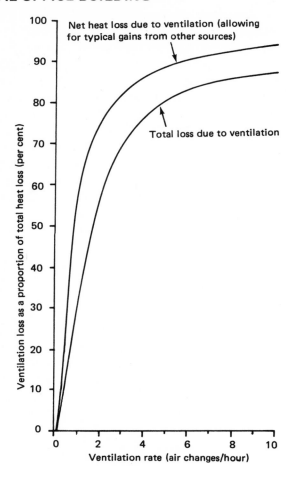

FIGURE 5.6: THE IMPORTANCE OF INTERNAL HEAT GAINS ON REQUIRED HEAT INPUT

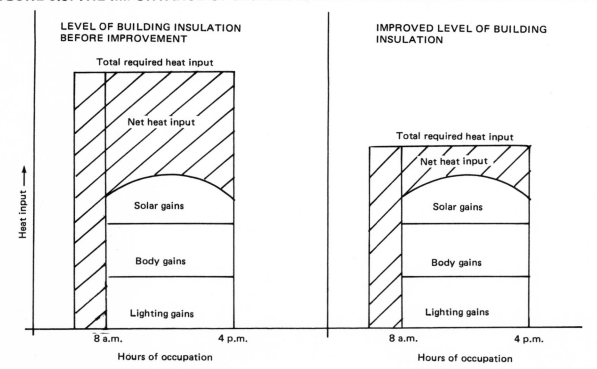

Pattern of Use

The majority of buildings can be operated between two temperature levels during the heating season. The upper level is the normal occupation temperature, and the lower level is either an anti-condensation temperature or a minimum comfort level e.g. for sleeping purposes in a hospital or in elderly persons' homes. In order to conserve energy, the buildings should be operated with the minimum pre-heat period to attain the upper temperature level, and the heating switched off until the lower level is reached by the natural cooling of the building. Energy will be wasted if the Night Set Back flow temperature is above the minimum level required by the heating plant because any temperature maintained above the minimum level during the unoccupied period causes a greater heat loss than necessary from the building.

The use of an optimum start controller is to be recommended to replace any simple time switch controller. This will allow the heating plant start-up time to be delayed during mild weather conditions rather than start the plant at some fixed point in time as happens with a normal time switch. Savings of between 10 and 25 per cent of the existing energy consumption can be made depending on the type of building, type of heating system, and period of occupation.

Comfort Level

It is necessary to understand the physical environmental factors which influence the thermal comfort of building occupants. Figure 5.7 summarises the relationships between activity and body response for given globe thermometer temperature levels. (Globe thermometer readings differ from conventional thermometer readings in that they measure the absorption of radiant energy). For thermal comfort to be maintained the human body must lose amounts of heat proportional to the amount of physical activity. This heat can be lost in a number of ways and to a limited degree the balance between the different ways can be varied, such as by choice of clothing.

FIGURE 5.7: RELATIONSHIP BETWEEN ACTIVITY, GLOBE THERMOMETER TEMPERATURE AND COMFORT

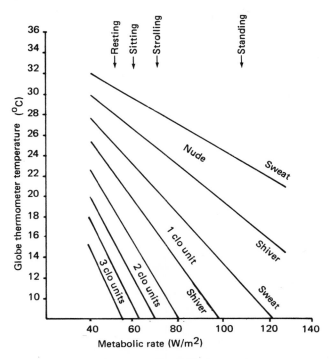

A globe thermometer is a thermometer with its bulb at the centre of a 150 mm diameter matt black sphere.

FIGURE 5.8: FACTORS IN BUILDING HEAT BALANCE

Initial Conditions	Heat Gains	Heat Losses	Satisfactory Comfort Conditions
Thermal properties of the building Absolute humidity of the air (unless air conditioning with humidity control is provided)	Solar radiation Heat from occupants Lighting Mechanical installations and equipment Heating installation	Radiation to sky Convection to air outside Ventilation losses into ground Refrigeration plant	Mean radiant temperature Air temperature Relative humidity (A satisfactory rate of air movement must also be given by ventilation arrangements)

People can tolerate a much wider variation of environmental conditions when active than when at rest. Thermal and thermally related factors which are at work inside buildings must be kept within a relatively narrow range of balance if comfort is to be achieved. Figure 5.8 gives an indication of the factors involved in achieving this balance.

To make valid thermal predictions and comparisons, it has been necessary to make some qualifications of the thermal effect of clothing. Scales were developed in both England and America; the American scale of units called clo-values is the one which has gained widest acceptance and use. The scale varies from zero for no clothes at all through one clo-unit which represents a normal suit and underwear, up to a maximum of about four which represents heavy polar dress. The unit is scientifically defined in terms of heat transfer resistance from the skin to the outer surface of the clothed body.

Table 5.9 shows a typical range of combinations of clothing together with their appropriate clo-value, and typical temperatures at which sedentary subjects would be thermally comfortable. It will be observed that comparatively modest variations in clothing have marked effect on comfort temperatures and consequently upon energy conservation in buildings.

Insulation is of great importance in establishing comfort conditions as they are affected significantly by the surface temperature surrounding the building occupants. The value of high levels of insulation can be readily seen by examining Figure 5.10.

Efficiency of Heat Utilisation

The efficiency of heat utilisation within a building is largely a measure of the effectiveness of the heat distribution system. This is a subject in its own right (see Chapter 6, Energy Managers Workbook Volume 1). Efficiencies of less than 50 per cent are by no means unknown, particularly in factories.

Efficiency of Heat Generation

Under normal circumstances, the efficiency of heat generation for use in buildings would vary somewhat between 75 and 85 per cent, although in some instances poor load factors lead to thermal efficiencies below 60 per cent.

Heat generation is also a subject in its own right and can only be touched upon in this chapter.

TABLE 5.9: THERMAL EFFECT OF CLOTHING

Clo value		Sedentary and resting max comfort temp °C
0	Nude	28.5°
0.5	Short underwear Light cotton trousers Short-sleeved, open-neck shirt	25.0°
1.0	Short underwear Typical business suit, including a waistcoat	22.0°
1.5	Long underwear Heavy tweed business suit and waistcoat Woollen socks,	18.0°
2.0	Long underwear Heavy tweed suit and waistcoat Woollen socks, heavy shoes Heavy woollen overcoat Gloves, hat	14.5°

FIGURE 5.10: THE IMPORTANCE OF INSULATION FOR COMFORT CONDITIONS

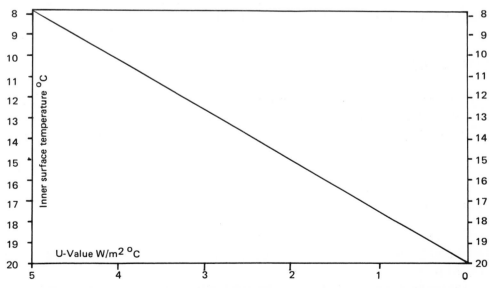

Relationship between U-value and inner surface temperatures for normal exposure and with a 20°C internal/external temperature difference. The figures at the extreme ends of scales are the theoretical maximum and minimum design values.

TABLE 5.11: FACTORS TO BE CONSIDERED BY THE ENERGY MANAGER

Use	Encourage multiple use	In appropriate circumstances (particularly educational buildings) multiple use will give better economy of operation and may liberate whole buildings.
	Reduce comfort temperature	Inevitably associated with wearing more clothes. One additional pullover can have a significant effect. Present clothing levels are made possible by cheap energy. Increased energy costs will lead to a new balance of clothing and heating.
Siting	Avoid exposed sites	Increased energy requirements for buildings on an exposed site are mainly due to higher wind speeds giving reduced U-values and increased ventilation rates. Elevated sites can be colder and more subject to rain and mist.
	Avoid noisy and polluted sites	Where excessive noise or atmospheric pollution is present, windows cannot be opened and mechanised ventilation (and probably air conditioning) will be required.
Planning	Reduce volume	For a given occupancy, the smaller the volume of the building which can be achieved, the better the energy economy. Loss through the fabric decreases when the area of external surface is reduced and the ventilation loss is diminished also. The heat gain from lighting, occupants and processes will, in a smaller building, contribute a much greater proportion of the heat required. Apart from operational savings, there is a substantial saving in the energy required for basic construction.
	Use economical shapes	Cubical shapes with flat facades have smaller surface areas and consequently smaller heat losses than elongated or elaborately configured shapes.
	Group individual buildings together	Small individual buildings will have greater heat losses than the same buildings combined into a single block. Semi-detached houses have lower heat losses than detached, and terraces are better than semi-detached.
	Avoid open plan	Open plan buildings mean that the whole volume must be heated, whereas with separate rooms differential control of temperature is possible, and ventilation losses may be reduced.
	Locate heat sources centrally	Boilers and flues and any other heat source should be located centrally so that heat is contributed to the other rooms in the dwelling and not lost to the exterior. (Often with unfortunate consequences, as in flue condensation).

Table 5.11: Factors to be Considered by the Energy Manager (Cont)

Construction	Provide adequate insulation	Structural insulation not only reduces heat losses but also raises the internal surface temperatures which will improve comfort and may even permit the use of lower air temperatures.
	Provide appropriate thermal response	If completely continuous heating is to be provided, no special consideration of the response of the structure is needed. Except for hospitals and prisons, however, few buildings have continuous heating. High thermal capacity walls and floors lose heat during the period when the heating is off and require substantial pre-heating periods to come back to comfort conditions when the heating is turned on. Low thermal capacity linings are economical of energy in most cases and critically important when heating is very intermittent.
Fenestration	Reduce window sizes	Even double glazing has three times the rate of heat loss of an external wall. To conserve energy, window areas should be kept to a minimum. Curtains or shutters for use at night cut down heat losses at the most critical times.
	Consider window orientation	Solar heat gain through windows is a significant factor which varies considerably with orientation. Unobstructed, south facing windows, curtained at night, gain as much solar heat as they lose during the winter. East and west windows gain some heat, north windows very little.
Energy source	Select energy conserving fuel	The availability of fuel and the suitability of particular types for the building should be considered, together with the related primary energy requirements.
	Utilise waste or environmental heat	In appropriate circumstances, not usually domestic, various heat recovery measures can be taken to remove waste heat from exhaust air or flue gases and make it available for heating. Heat pumps may extract heat from the surrounding environment and thus provide economical energy.

Conclusions

The studies described have been concerned with the nature and scale of possible energy savings in the heating of buildings. The appropriateness of any particular method, the actual energy saving, the capital (and energy) investment required, and the choice must depend upon the circumstances of particular cases. It is clearly desirable, however, to attempt to draw some conclusions about the relative values of various methods of energy conservation, which are the responsibility of the designer or building manager.

In a number of instances direct generalised comparison can be made. In others, such as solar gain, direct comparison could only be made in the context of a specific problem. The case for attention to shape and solar gains is, however, entirely clear. In all new buildings careful thought should be given to shape, orientation and fenestration so that, within the limits of acceptable design, the best arrangement is made to minimise heat losses and to maximise gains during the heating season. Summer overheating must be checked and avoided.

Except in housing, where these considerations are an important factor, the potential energy gains are not that substantial, but since no significant cost or other penalty is involved, the return upon investment is high. Architects should therefore take this into account in design.

Shape and insulation are considerations for new buildings. In both new and existing buildings adjustments can be made to the thermal response of the buildings—mainly insulation and its position in walls and roofs—and to the control of the heating installation. Poor thermostatic control can give rise to substantial waste. In domestic buildings appropriate controls are not expensive relative to the increase in comfort and economy, and are in widespread use.

Great savings can be achieved by time switching the heating installation to operate only when really required. Also significant is the effect of improving the speed of thermal response of the fabric. Increasing insulation may be worthwhile since even small savings are important, but it is not highly effective in all situations studied. An exception is the case of continuous heating in constructions of very low basic insulation. Apart from hospitals and similar buildings there seems every reason to outlaw continuous heating.

Fortunately measures for energy saving are not usually mutually exclusive but it does appear that the most effective building measure is the time switch, followed by thermostatic control and fast response for the building itself and the heating installation.

There are important factors in energy conservation which are quite outside the control of the architect but which are nevertheless very significant to considerations of building design. Probably the most significant of these is the cost constraint on heating which is very real for many people. Studies at the Building Research Establishment (BRE) have shown that a large percentage of domestic users govern their heat consumption not by the thermostat but by the meter. If additional insulation is used in these cases, the result is likely to be improved comfort rather than energy saving.

Other aspects such as multiple occupancy and clothing will also influence energy needs and Table 5.11 highlights the types of considerations the energy manager should take into account within his workplace.

6

ASSESSING THE SCOPE FOR HEAT RECOVERY

Glenn Brookes

Of all the energy saving measures available, by far the greatest potential for saving exists through heat recovery. The reasons are not difficult to assess. The main industrial and commercial energy demand is for energy in the form of heat. Also, although these energy users are referred to as consumers, few processes actually consume heat energy; they merely degrade heat, reducing its temperature until it is finally rejected to the environment as warm air or water.

In the end, all energy used has to be rejected as heat. What is important is that the best possible use is made of it and that fuel for heating is not purchased where it can be obtained more economically by recovering heat that would otherwise be rejected.

Before looking at methods of heat recovery in detail it is worthwhile to examine the relationship between heat and temperature.

When we add heat to, or remove heat from, a substance, its temperature changes. The amount of temperature change is related to the quantity of heat through the specific heat (sometimes called the heat capacity) by:

quantity of heat = temperature change x specific heat x weight of material

The specific heat is a property of the material and is usually expressed in units of energy per unit weight per degree (e.g. $kJ/kg/^{o}C$ or $Btu/lb/^{o}F$). When dealing with heat in a gas such as air it is more appropriate to use volume, so that:

quantity of heat = temperature change x volume specific heat x volume

In heat recovery the main concern is the rate of heat production or transfer. This is represented as:

rate of heat production (or transfer) = temperature change x specific heat x mass flow rate

or

rate of heat production (or transfer) = temperature change x volume specific heat x volume
flow rate

Rate of heat production can then be expressed as kilowatts, therms per hour etc.

There are actually two kinds of heat: sensible heat and latent heat. Sensible heat relates to the 'hotness' of a substance and is lost or gained by changes in temperature, whereas latent heat is heat absorbed or released by a substance without a temperature change. This is usually associated with a change of phase of a substance, such as the melting or freezing of ice, or the evaporation or condensation of steam or water vapour. Indeed, in practical heat recovery problems latent heat is usually encountered with heat recovery either from steam, which is simple to calculate, or from moist air (in air conditioned buildings or dryers), which involves more complicated calculations using psychrometry.

Identifying Opportunities

It is necessary first of all to identify all the significant places in the factory or building where heat is being lost, and then measure the temperature and the mass or volume flow rate. Some common sources of waste heat are given in Table 6.1. Solid materials are always cooled by some secondary medium such as water or air and it is the temperature and flow rate of this medium that concerns us most.

TABLE 6.1: SOME COMMON SOURCES OF WASTE HEAT

Solid	Liquid	Gas
Fired bricks	Process cooling water	Dryer exhausts
Fired ceramics	Air compressor cooling water	Burner flues
Cooked foodstuffs	Refrigeration cooling water	Boiler flues
Metal products	Engine cooling water	Flash steam
Chemical products	Condensate	Kiln and oven exhausts
	Liquid process streams	Building ventilation
	Boiler blowdown	Engine exhausts

The amount of heat which can be recovered from these streams then depends on the extent to which their temperature can be reduced; this in turn depends on where it is going to be used. For example, if warm air from one place in the factory is used to preheat ambient air somewhere else, then in principle the air could be cooled to the outside air temperature. If, however, it is a water stream at 100°C being used to heat a process stream which is already at 55°C, the maximum heat recovery corresponds to the difference in the sensible heat content of the water stream at 100°C and 55°C.

So, the amount of heat available cannot be determined until its use is known. The next step, therefore, is to identify the places where recovered heat may be used, the temperature at which it is required, the temperature changes which the heat is used to effect in these applications and the amount of heat they are currently using. With this information it should be possible to make a preliminary match of heat sources to possible uses.

Where to Use Recovered Heat

When the range of opportunities available for matching heat sources to possible applications are examined, four classes of match emerge. These arise from the fact that recovered heat can be used directly or indirectly, i.e. with or without the use of heat exchangers, and can be used either in the same process from which it was recovered or in a different one. The four classes are:

  —direct use in the same process
  —indirect use in the same process
  —direct use in a different process
  —indirect use in a different process.

These have been listed in what is generally the order of decreasing attractiveness.

Broadly speaking, whether the heat can be used directly depends on the quality of the medium in which it is carried. One prime consideration is whether there are contaminants or particulate material in the stream, e.g. fume, odour or dust in air, or dissolved solids or scale in water. Some applications, such as ventilation air in buildings, air for dryers, washing water and boiler feed water, have very strict quality requirements.

When heat can be re-used in the same process it offers two attractions. One concerns timing. If the heat can be returned to the process from which it came, the availability of heat and the need for heat are naturally synchronised thus avoiding the need for expensive heat storage techniques. If heat is transferred to another process it is likely that there will be times when it will be needed but will not be available, and will have to be made up from elsewhere, or it will be available but not required, and will therefore be wasted.

The other advantage concerns distance. It is costly to move heat around. Usually, heat will need to be transported further if it is used in a different process. In this context it should be noted that because of the low density of air it is considerably more expensive to move heat in the form of hot air or gases than as a liquid.

Direct Use in the Same Process

Examples of direct use of recovered heat in the same process are partial recycle of exhaust air in dryers, destratification of rising air in buildings and return of condensate to boilers. So far as process plant is concerned, retrofit opportunities to recover heat directly and use it in this way are becoming less common, as newer plant tends to be designed with these features already incorporated. In older plant there may still be opportunities for direct use of recovered heat. Furthermore, because such measures are low cost, they are often worthwhile even in plant with a short residual life.

Indirect Use in the Same Process

By far the greatest number of heat recovery opportunities are in this class. Examples are vapour recompression in evaporator plant, heat recovery between burner flues and process air in ovens and kilns, heat recovery between exhaust air and burners (recuperative burners) in kilns and furnaces, flue gas economisers in boilers, outlet to inlet heat recovery in dryers, building heating air, pasteurisers and sterilisers. Again, some of these heat recovery systems are increasingly being incorporated into new plant and buildings.

Direct Use in a Different Process

Though often a low cost option, this is not common because of quality and timing considerations. Examples include the use of evaporator condensate for washing, and the use of furnace or kiln exhausts for drying applications.

Indirect Use in a Different Process

These opportunities are much more widespread. Examples of common sources of heat are refrigeration plant, air compressors, generators, boiler flues, dryer, oven and kiln exhausts, burner flues, process streams, cooling towers, flash steam etc. Applications include space heating, air preheating for processes, feedstock and raw material preheating, water heating for processes, domestic hot water heating, boiler feedwater heating etc.

Assessing the Case

First Steps

The first step in assessing the case for heat recovery is to identify the main uses of energy on the site. This gives an indication both of places where recovered heat might be used, and also, on the principle that energy rarely gets significantly consumed, where heat is being rejected. For an initial impression of the amount of heat available, estimating the heat going into a process is often easier than trying to measure what comes out, and it is even possible to meter it.

A list should now be drawn up of all the available waste heat streams, together with temperatures, flow rates, pressures, amounts of latent heat, and quality of the stream. The timing and location of each of these should be indicated.

With this information it should be possible to begin shortlisting potential matches between available heat and end uses.

Proceeding with the Assessment

Before going into any detailed assessment two important questions must be raised:

    —is it necessary to reject this much heat anyway?
    —is it necessary to use heat in the proposed application?

These ought to be obvious. If a process which uses heat is working inefficiently the excess heat it uses will appear in the reject heat. It is important to ensure, therefore, that the process is working efficiently before heat recovery is considered. If a process is not running properly, other problems can arise. Take the example of heat recovery from the flue of an oil-fired process. If the burner is worn or damaged and cannot be set up correctly, it will operate inefficiently with a sooty flame that can rapidly foul a heat exchanger and reduce its efficiency or even block it completely.

Assessing Financial Viability

Once the heat stream from which heat is to be recovered and the use to which it is going to be put are known, the amount of heat to be recovered can be determined. If the sort of equipment to use is an unknown factor the costs cannot yet be calculated, but knowing the amount of heat and the cost of heat can enable a first estimate of the value of the heat to be made.

Suppose there is a source of heat at 85°C from which 50 kW of heat can be recovered for six hours a day for five days a week, 49 weeks a year. The total heat recoverable in a year is:

50 x 6 x 5 x 49 = 73,500 kWh per annum

Suppose this heat is now used to heat water at 10°C to 65°C, where previously it was heated by steam provided by a boiler burning medium fuel oil at 17 pence per litre. For medium fuel oil 17 pence per litre is equivalent to 1.515 pence per kWh. If the boiler converts at 75 per cent efficiency the cost of heat output from the boiler is:

1.515 ÷ 0.75 = 2.02 pence per kWh

The value of the recovered heat is therefore:

£2.02 ÷ 100 x 73,500 = £1,485 per annum

If the maximum payback requirements are three years then equipment that costs up to £4,455 can be considered.

A wise precaution at this stage is to test the effect on the value of heat of any assumptions already made. In this case, there are four assumptions about temperature—that heat is available at 85°C, that it is used to heat water to 65°C, that the water intake temperature is 10°C and that heat can be transferred from 85°C to 65°C in a heat exchanger.

The sensitivity of the cost calculation to errors in the temperatures should be examined. In this case, for every 1° error in either the water intake temperature or use temperature the calculated heat delivered will be in error by

$$\frac{100 \times 1}{(65 - 10)} = 1.8 \text{ per cent}$$

Clearly, as the temperature fall in the waste heat stream or the temperature rise in the stream to which the heat is delivered become smaller, errors in these temperatures will become more significant. Assumptions about temperatures become especially important when heat pumps are under consideration.

The temperature difference in a heat exchanger is important because it affects its size, and hence its cost. The size of a heat exchanger is given by:

heat recovered = area x constant x ΔT

where ΔT is the difference between the inlet temperature of the hot stream and the outlet temperature of the stream being heated. It is important to keep this difference as large as possible to eliminate errors. For a given quantity of heat recovered area is related to this temperature difference as shown in Figure 6.2.

**FIGURE 6.2: RELATIONSHIP BETWEEN CAPITAL COST AND TEMPERATURE
DIFFERENCE FOR HEAT EXCHANGERS**

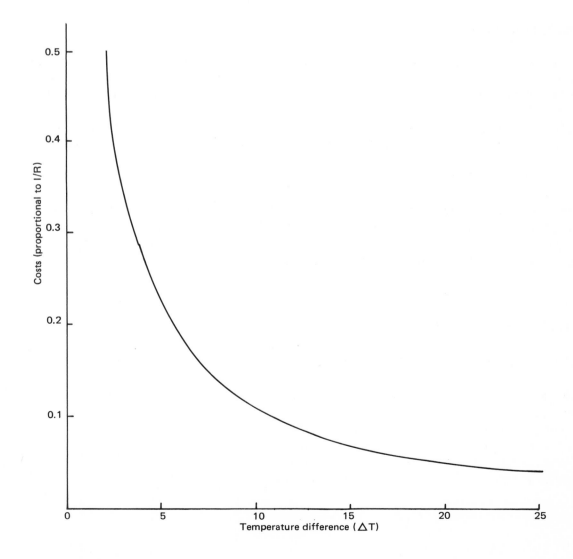

Heat Recovery Equipment

Although there appears to be a very wide choice of energy saving equipment available, in practice the particular requirements of a heat recovery application usually limit the choice to one or two types of system.

The first main division depends on whether the waste heat stream and the application involve heat exchange with gas or a liquid. This gives rise to three classes of equipment—gas/gas, gas/liquid, liquid/liquid. The main types of heat recovery equipment in each class are indicated in Table 6.3.

TABLE 6.3: THE MAIN CLASSES OF HEAT RECOVERY EQUIPMENT

Gas/gas	*Gas/liquid*	*Liquid/liquid*
Rotating regenerator	Economiser	Shell and tube heat exchanger
Plate heat exchanger	Waste heat boiler	Plate heat exchanger
Run-around coil	Fluid bed heat exchanger	Spiral heat exchanger
Heat pipe heat exchanger	Spray recuperator	Heat pump
Static regenerator		
Tubular (convection) recuperator		
Tubular (radiation) recuperator		
Heat pump (as dehumidifier)		
Recuperative burners		
Regenerative burners		
Glass tube heat exchanger		

GAS/GAS HEAT RECOVERY EQUIPMENT

Rotating Regenerator (Heat Wheel)

The rotating regenerator consists of an open structured matrix of metal in the form of a wheel which rotates and thereby moves segments of the wheel alternately between the two air streams. (See Figure 6.4). The material in the matrix can be any open structured metal—knitted wire, honeycomb or corrugated sheet—and can be made of fibrous, steel, aluminium or ceramic material. Ceramics are used for very high temperature applications.

FIGURE 6.4: SCHEMATIC ILLUSTRATION OF THE OPERATION OF A HEAT WHEEL

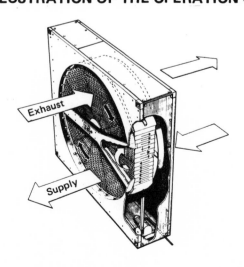

By using an etched surface or a surface coated with a hygroscopic material in the matrix, the regenerator can be used to pick up the latent heat from moist air. This makes it particularly useful for air conditioning applications and heat recovery in high moisture environments such as swimming pools.

Merits: High temperature efficiencies possible; can recover latent heat; available in a range of sizes; wide operating temperature range; low pressure drops; possible to replace core matrix.

Limitations: Cross contamination; moving parts; tend to be bulky; can tolerate only small pressure differentials between gas streams.

Plate Heat Exchanger

In its simplest form a plate heat exchanger consists of a framework of parallel metal plates, separated from each other and sealed at the ends in such a way that the exhaust and inlet gases flow through adjacent passages with heat being transferred between them by conduction through the separating metal sheet. (Figure 6.5).

FIGURE 6.5: SCHEMATIC ILLUSTRATION OF THE OPERATION OF A SIMPLE PLATE HEAT EXCHANGER

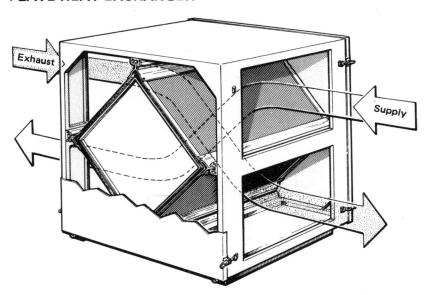

The gas flows can be modified to provide essentially counter flow movement of inlet and exhaust streams, thereby improving the heat transfer performance of the exchanger. Improvements can also be brought about by corrugating the metal sheets or adding finning to the metal surfaces.

The heat exchanger is typically aluminium for low to medium temperature applications, but all welded stainless steel units are available for high temperature work.

Merits: No cross contamination; easy to install on-line cleaning; can operate under condensing conditions; simple construction.

Limitations: Only small temperature differentials can be tolerated with some designs; large, inconvenient ducting arrangement in some circumstances.

Tubular (Convection) Recuperators

The materials and methods of construction of tubular convection recuperators are mainly influenced by the operating temperature and nature of the hot gas stream. The basic design consists of a bundle of tubes, through which passes one gas stream, and over which passes the other. The pass over the tubes can be single cross flow, but to improve heat transfer a number of passes can be arranged.

Metallic recuperators, using high nickel/chrome steels, are normally used for gas temperatures up to 1,000°C. For temperatures below 400°C, glass tube heat exchangers, as illustrated in Figure 6.6, are being increasingly used, particularly when the exhaust gases are corrosive and the unit is operating under condensing conditions.

FIGURE 6.6: SCHEMATIC ILLUSTRATION OF THE OPERATION OF A GLASS TUBE HEAT EXCHANGER

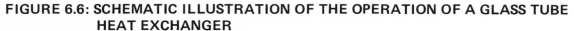
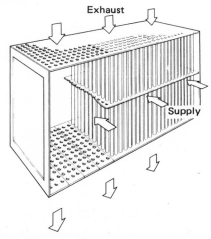

The tubes are normally plain bore with no extended surface. This means that more tubes are needed to provide the necessary surface area for heat transfer, but they have the advantage of being less susceptible to fouling.

Merits: Corrosion resistant; easy to clean; made to any size; can see inside; no cross contamination.

Limitations: Limited temperature range; bulky.

Run-Around Coil

The devices described so far require the ducts carrying the gases to be brought to a common heat exchanger. This usually involves the construction of ducting. Above a certain distance it is far cheaper to move heat as a hot liquid in a pipe. This is the principle of the run-around coil (Figure 6.7).

The run-around coil uses two gas/liquid heat exchangers (which is more expensive than using just one exchanger) connected by a pipe containing a heat carrying fluid (which is cheaper than ducting). Usually the fluid is water, sometimes containing glycol anti-freeze, although refrigerants have also been used.

In terms of overall cost, when distances are very short gas/gas exchangers are the most economic; when distances are great the run-around coil has more advantages. The choice is a fine balance in which all factors need to be taken into account, including:

—cost of the fluid and its possible replacement
—cost of pumping
—effect of anti-freeze on pumping cost
—cost of supports for ducts or pipes.

FIGURE 6.7: SCHEMATIC ILLUSTRATION OF THE OPERATION OF A RUN-AROUND COIL SYSTEM

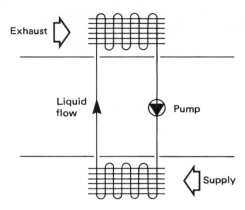

Merits: Can economically cover large distances; easily tailor made; can accommodate multiple heat sources and applications; can match materials of construction to source and sink; easily controllable.

Limitations: Generally lower temperature efficiency; moving parts; temperature range limited.

GAS/LIQUID HEAT RECOVERY EQUIPMENT

Economisers

These are used extensively for waste heat recovery from boiler exhausts and generally consist of a bundle of finned tubes spanning an exhaust duct. The liquid to be heated is pumped through the tube bundle, usually making several passes of the gas stream.

Economisers are commonly made from cast iron or steel, and are made substantial enough to cope with the aggressive environment of a boiler flue, particularly with respect to back end corrosion caused by condensation of water. An economiser can be constructed from more exotic materials such as plastic, and the fluid to be heated could be an organic high temperature heat transfer fluid.

Merits: Well developed; easily tailor made; variety of materials of construction.

Limitations: Can suffer from cold spots leading to corrosion.

Waste Heat Boiler/Shell and Tube Heat Exchangers

The design difference between conventional and waste heat boilers arises largely because of the greater variety of environments in which the latter have to operate. A shell and tube heat exchanger is a variant on a waste heat boiler in that the principles of operation are the same, but the liquid stream does not vaporise. For the smaller heat recovery duties the gas stream generally passes through the tubes, but for larger duties and higher operating pressures (greater than 200,000 kg/hr steam and greater than 20 bar pressure) water tube units are the norm.

Because of the high heat transfer rates associated with a waste heat boiler, the units tend to be more compact and lower on capital cost than other heat recovery systems of comparable duty.

Merits: Compact, high heat transfers; easily cleanable; can operate in fouling environments.

Limitations: Not generally internal to process; must link with steam distribution network.

Fluidised Bed Heat Exchanger

Fluid bed technology, although developed mainly for high efficiency combustion, can be applied to waste heat recovery. It can be particularly useful when the hot waste gas stream is heavily contaminated.

Typically, flue gases are diverted from a main stack to a shallow fluidised bed, where finned tubes located in the bed pick up heat in a circulating flow of hot water, steam or an organic thermal fluid. Gas temperatures up to $1,000^{\circ}$C can be accepted.

Merits: Compact; can deal with foul streams; self-cleaning.

Limitations: Early stage of development; moving components.

Direct Contact Heat Exchangers/Spray Recuperators

As the name implies, this method of heat recovery involves direct contact between the waste gas stream and the liquid to be heated. Commercially available spray recuperators spray water into a hot gas stream, usually a boiler exhaust, which results in the recovery of both sensible and latent heat. The hot water is then collected in a holding tank and may be pumped to a liquid/liquid or liquid/air heat exchanger as required.

Merits: Good heat transfer; simple; recovers latent heat and water.

Limitations: Lower grade heat; less appropriate to oil-fired plant.

LIQUID/LIQUID HEAT RECOVERY EQUIPMENT

Shell and Tube Heat Exchangers

The shell and tube heat exchanger is well established and consists of a number of tubes spaced a small distance apart, with one fluid flowing through the tubes and the other over the tubes (the shell side).

Variations on this design are possible and include the use of baffles to act as flow deflectors and a floating tube head to permit the use of two tube side passes and allow the tube bundle to be withdrawn.

Merits: Well developed; off-the-shelf ranges available; comparatively cheap; usable at high pressures.

Limitations: Often poor flow characteristics; fixed duty.

Plate Heat Exchangers

The standard plate heat exchanger consists of a frame in which metal plates are clamped together and separated from each other by corrugations of the plate and gasket seals around the periphery. The liquid flows through the exchanger are arranged to ensure that the two liquids flow counter current to each other between adjacent parallel passages. (Figure 6.8).

Very high thermal efficiencies can be achieved with these exchangers with a unit of acceptable size and capital cost. They can be easily dismantled for cleaning and it is a simple matter to increase the number of plates to increase the heat exchange duty.

FIGURE 6.8: SCHEMATIC ILLUSTRATION OF THE OPERATION OF A PLATE HEAT EXCHANGER

Merits: Easy to clean; compact; variety of materials of construction; flexible duty.

Limitations: Maximum temperature set by gasket materials; low pressure use only.

Other Specialist Equipment

We have covered here only a selection of the most common heat exchanger types. There are other types for special circumstances including:

—spiral heat exchangers for high pressure applications
—heat pipe heat exchangers for compact air to air duty
—demountable shell exchangers for use with heavily fouling liquids
—radiant tube recuperators for high temperature gas/gas duty
—heat pumps including a wide range of equipment for use where the temperature of waste heat needs to be increased
—vapour compression for raising the quality of recovered steam.

7

PROCESS INTEGRATION

Greg Ashton

In the previous chapter the identification of heat recovery opportunities and the selection of appropriate equipment was discussed. It was assumed for the most part that a single source of heat might be matched to a single application. However, the range of opportunities for heat recovery is often more complex. For example, because of flow rate, specific heat or temperature considerations the supply of heat to an application might require heat recovery from more than one source. Alternatively, after taking heat from a waste heat stream there may be further heat available at a lower temperature to enable an additional use elsewhere.

Following through this idea leads directly to the concept of heat recovery networks, i.e. multiple exchanges between many streams. Heat recovery networks are common in large scale power plant such as chemical plant, power stations and refineries, and are also found in small production plants.

A major problem with the design of heat exchanger networks until recently was how to determine the optimum arrangement of heat exchangers for maximum heat recovery and minimum cost. A considerable breakthrough was made with the discovery of a number of basic rules of thermodynamics which greatly simplify the problem. These rules, and the methods of applying them, have been developed entirely in the last ten years by Professor Bodo Linnhoff of the University of Manchester Institute of Science and Technology (UMIST) and have been applied to a wide range of processes.

The techniques are applicable to many problems other than those of large plant, including the design of combined heat and power systems and optimum use of heat pumps. They can also be used to enable improvements to be made to plant which is less complex than refineries but which is still difficult to handle by other methods. It is in these areas that the techniques become of interest to most energy managers.

It is not essential for the average energy manager to be able to use the techniques himself, but it is important for him to be aware of what the techniques are and what can be achieved with them.

Basic Principles

Figure 7.1 shows a flowscheme for a simple continuous process in which feed stream 1 is to be cooled and product streams 2 and 3 are to be heated. The heat exchanger network shown requires 160 units of utility heating by steam and 70 units of utility cooling from cooling water, assuming a $20^{o}C$ minimum temperature approach between streams.

The network shown in Figure 7.1 is certainly feasible, but it can be improved by designing a maximum energy recovery network, for which purpose the basic principles of process integration need to be known.

FIGURE 7.1: A SIMPLE HEAT EXCHANGER NETWORK

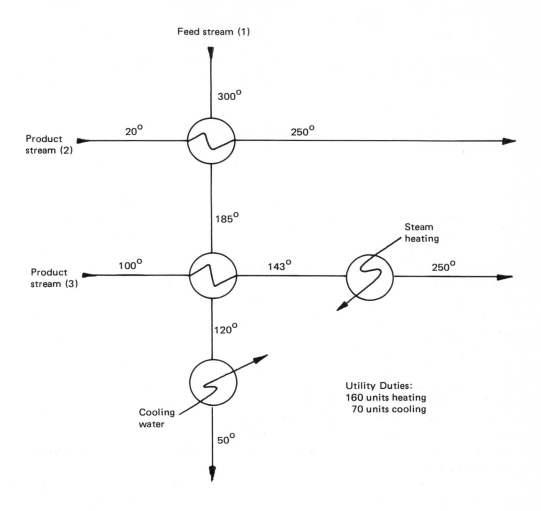

The first step in analysing any problem in terms of process integration is to construct a table in which all the process streams are listed and classified according to whether they give up heat (hot stream) or receive heat (cold stream). Each stream then needs to be specified by flowrate, specific heat and temperature information. For example, the stream data for the process in Figure 7.1 is set out in Table 7.2. The product of the flowrate (F) and specific heat (Cp) is given by the symbol CP.

TABLE 7.2: STREAM DATA FOR THE PROCESS SHOWN IN FIGURE 7.1

Stream No	Type	Stream supply temperature $^{o}C_1$	Stream target temperature $^{o}C_2$	Flowrate (F)	Specific Heat (Cp)	CP = (F) x (Cp)
1	Hot	300	50	2.0	0.5	1.0
2	Cold	20	250	1.0	0.5	0.5
3	Cold	100	250	1.5	1.0	1.5

1. Stream supply temperature = temperature at which the stream enters the process
2. Stream target temperature = temperature to which the stream must be heated or cooled

The data in Table 7.2 can now be used to represent the process streams on a temperature vs heat diagram, as shown in Figure 7.3. In general, any stream with constant CP would be represented on this diagram by a straight line running from stream supply temperature to stream target temperature. Thus, in Figure 7.3 the hot stream (1) is represented in this way by the line A-B. However, the lines for the cold streams (2, 3), have been combined by totalling the heat contributions in the appropriate temperature intervals and then plotting the resultant line CDE, which is known as the cold composite curve. A similar summing procedure would have been used to produce the hot composite curve if there had been more than one hot stream in the data table.

FIGURE 7.3: COMPOSITE CURVES FOR THE STREAM DATA IN TABLE 7.2

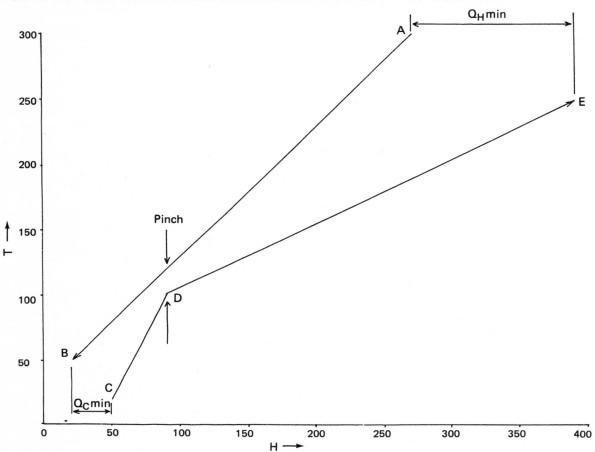

The composite curves shown in Figure 7.3 have been positioned so that the hot composite curve is hotter than the cold composite curve at every point, which means that heat transfer is possible. The minimum temperature approach between the two curves on any vertical axis is 20°C and the point where this occurs is defined as the 'pinch'.

The basis of the techniques of process integration lies in understanding the significance of the pinch. This is the point on the composite curves where the minimum temperature difference between hot and cold occurs. As the curves are moved relative to one another horizontally, the temperature difference at the pinch increases. At the same time the overlap—Q_H min and Q_C min in Figure 7.3, which are respectively the minimum amounts of external heating and external cooling required—also increases. If the curves are moved closer so that they actually touch, the overlap represents the absolute theoretical minimum of these values.

Furthermore, it can be shown that in order for the exchanger network to meet the targets represented by Q_H min and Q_C min it must obey the following rules:

—heat transfer from temperatures hotter than, to temperatures colder than the pinch must be zero
—external cooling should not be applied above the pinch
—external heating should not be applied below the pinch.

In the case of the process shown in Figure 7.1, the minimum utility targets are found to be 120 units of external heating and 30 units of external cooling. Therefore, there is maximum scope for 25 per cent steam saving, and in fact this saving could be realised by the heat exchanger network shown in Figure 7.4, which has been designed in accordance with the above rules for maximum energy recovery.

(For further details on the principles introduced above, including a fuller account of the network design methods used to generate Figure 7.4, 'A User Guide on Process Integration for the Efficient Use of Energy' published by the Institution of Chemical Engineers, is recommended reading.)

FIGURE 7.4: MAXIMUM ENERGY RECOVERY NETWORK GIVES 25 PER CENT STEAM SAVING AND MEETS TARGET

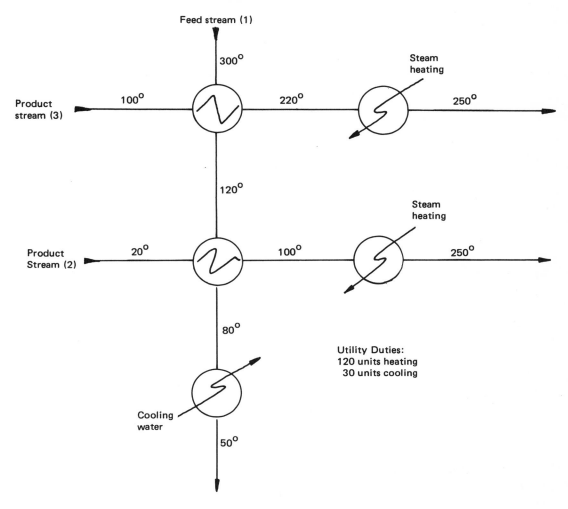

Balancing Energy Saving with Capital Costs

The composite curves in Figure 7.3 have a temperature difference of 20°C at the pinch, but what if the relative positions of the curves are moved horizontally so that the pinch temperature difference (ΔT min) changes? Figure 7.5 shows the same composite curves when the pinch temperature difference is 10°C. The energy targets have been reduced to 100 units of heating and 20 units cooling, but in order to realise this energy saving the heat exchanger network must have a larger total heat transfer surface area. This can be deduced from considering the general equation for heat transfer:

$$Q = U A \Delta T$$

FIGURE 7.5: REDUCED $\triangle T_{MIN}$ GIVES REDUCED ENERGY TARGETS

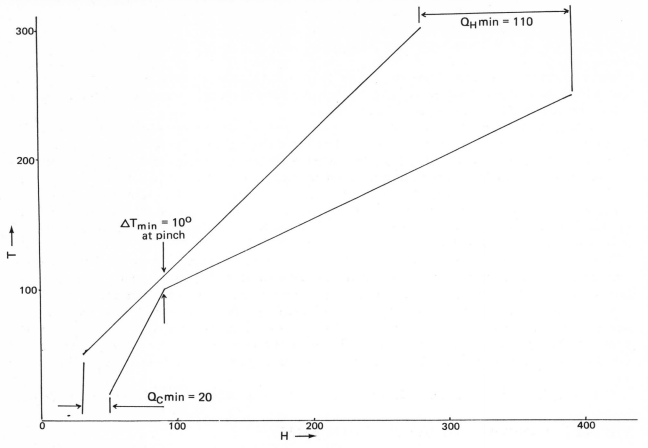

FIGURE 7.6: INCREASED $\triangle T_{MIN}$ GIVES INCREASED ENERGY TARGETS

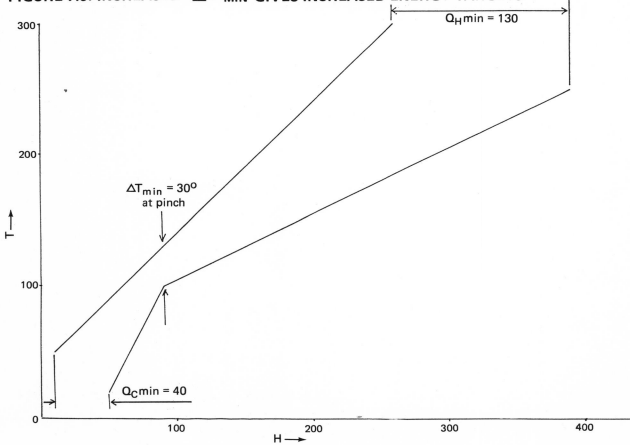

Where   Q = heat transferred
                 U = overall heat transfer coefficient
                 A = area of heat transfer surface
                 ΔT = mean temperature difference

Clearly, in order to transfer the same amount of heat at the same overall heat transfer coefficient but with a smaller temperature difference, then the heat exchanger area must be increased. The effect this has on costs was covered in Chapter 6.

Similarly, if the same composite curves are drawn with a pinch temperature difference of 30°C, (Figure 7.6), then the energy targets are increased to 130 units heating and 40 units cooling, but less heat transfer area is required. Obviously, the trade-off between energy and capital costs must be calculated in order to determine the optimum pinch temperature difference for design.

A modified form of the heat transfer equation can, in fact, be used to estimate the total heat transfer surface area for the heat exchanger network and the minimum number of heat exchangers required can also be calculated. These can, in turn, be used as the basis for a capital cost estimate against which energy costs can be set.

The latest development of process integration techniques at UMIST allows additional practical considerations to be taken into account in the capital cost estimate. These include varying stream heat transfer coefficients, materials of construction and heat exchanger configuration.

A convenient way of expressing the results of these calculations is to work out the combined yearly capital and energy costs and plot the total against pinch temperature difference. A typical graph of this type is shown in Figure 7.7. The discontinuities which often occur in these curves are due to changes in the number of heat exchangers required as pinch temperature difference varies.

This technique enables the calculation of optimum pinch temperature difference and therefore

FIGURE 7.7: CAPITAL—ENERGY COST TRADE-OFF

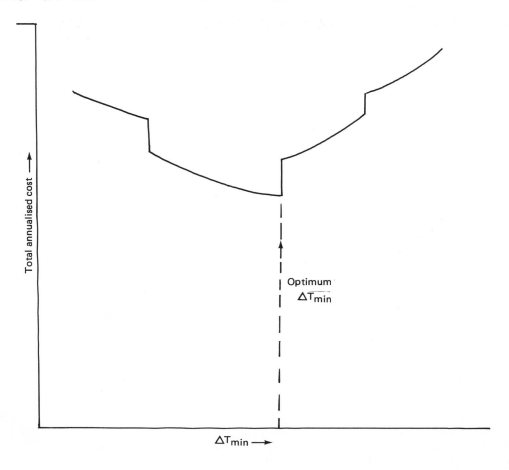

the energy targets for a heat exchanger network prior to design. Only the basic process stream data, a heat exchanger cost correlation and energy cost data have been used. No detailed design work regarding the choice of streams to be used for the exchange of heat has been carried out at this stage.

Identifying Energy Saving Process Changes

A more complex set of composite curves is shown in Figure 7.8, in which the hot composite curve contains a horizontal section, A-B, below the pinch. In practice, such a horizontal section would represent a hot stream condensing and giving up latent heat at a constant temperature—an aspect of heat recovery which presents no difficulty using these techniques.

If the pressure of this hot stream is increased so that its condensing temperature rises above the pinch, the composite curves will be modified to those shown in Figure 7.9 and the energy targets for both external heating and cooling will decrease significantly. If the required pressure increase is practical, then we have identified a simple process change which will save energy. This change has been revealed through the location and understanding of the pinch for this process and it is an aspect which any other approach to the problem is unlikely to expose.

The above example can be generalised and stated in the following terms. The energy targets decrease if:

- —hot stream duties are increased above the pinch
- —cold stream duties are decreased above the pinch
- —hot stream duties are decreased below the pinch
- —cold stream duties are increased below the pinch.

These statements are signified by plus and minus signs marked on the appropriate sections of the composite curves in Figure 7.9 and in the terminology used in process integration at UMIST this is known as the plus/minus principle.

Appropriate Integration of Heat Pumps

The plus/minus principle is most helpful in identifying energy saving process changes, but it can also be used to determine whether combined heat pumps are being used in an appropriate place in the process.

For example, consider the integration of a heat pump into the process described by the composite curves in Figure 7.9. An ideal heat pump takes in heat 'Q' at a low temperature (T_1), applies work (W), and rejects (Q + W) units of heat a higher temperature (T_2). The low temperature heat load of the heat pump therefore represents a new cold stream in the process, and if the temperature T_1, is below the pinch, then the corresponding duty, Q, is added to an appropriate section of the cold composite curve below the pinch.

Similarly, the heat rejected at the higher temperature represents a new hot stream with duty (Q + W). If the temperature T_2 also happens to be below the pinch, then this duty must be added to the hot composite curve below the pinch which, by the plus/minus principle, is inappropriate. In this case, in fact, the overall effect of incorporating the heat pump is to increase the external cooling target by 'W', which means that 'W' units of useful work have been taken and rejected as low grade heat into external cooling.

On the other hand, if the temperature T_2 is above the pinch, then the duty (Q + W) is added to the hot composite curve above the pinch, which is appropriate. The overall effect of the heat pump now is to reduce both the external cooling by 'Q', and the external heating by 'Q + W'.

The heat pump is now considered to be suitably placed. This simple rule for being able to determine how best to use a heat pump is an extremely important application of the techniques of process integration. Similar reasoning can also be applied to determine the appropriate placement of a heat engine, e.g. a gas turbine, showing that it should be placed entirely above or entirely below the pinch—a prediction which is otherwise difficult to make.

FIGURE 7.8: A PROCESS CHANGE SHIFTS THE CONDENSING STREAM FROM BELOW TO ABOVE THE PINCH

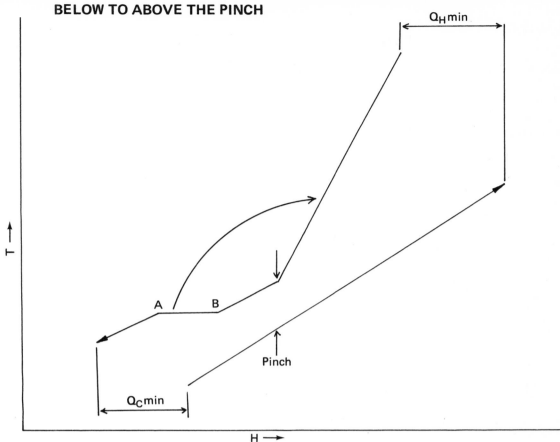

FIGURE 7.9: THE PLUS/MINUS PRINCIPLE

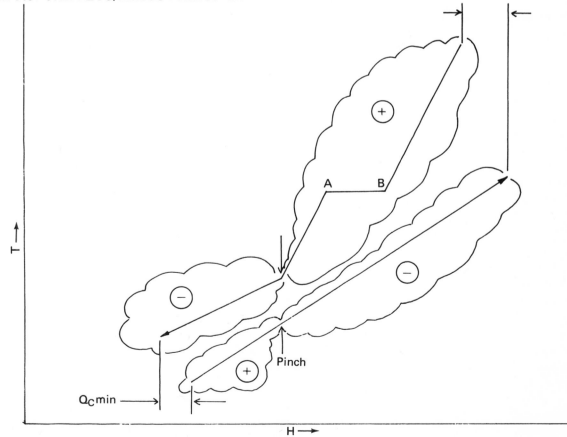

The Application of Process Integration Technology

The method used to apply process integration techniques in studies on existing sites can be considered to comprise three phases.

The first phase is the collection of data, which is preferably carried out by staff who are familiar with the plant, in close collaboration with a consultant. It is necessary to determine what information is required to compile stream data for the process and to measure as much as possible directly, often over a range of operating conditions. Practical limitations imposed, for example, by existing equipment, safety, operability and economic constraints should also be identified at this stage.

The second phase of the study can be called identification of possible opportunities. This would be carried out principally by a consultant who is familiar with the techniques of process integration. It will probably overlap with the first phase so that extra data collection can be carried out in parallel, as particular areas of the process are selected for closer evaluation.

The first step towards identifying energy saving opportunities is to set up the stream data for the process and construct the composite curves. This immediately indicates the minimum energy targets for external heating and cooling.

The composite curves can now be used to complement the intuition and experience of the engineer, in identifying specific improvements. This can be done firstly by inspecting the composite curves in the light of the plus/minus principle, in order to define process changes which will modify the composite curves so as to reduce the energy targets. Secondly, any heat exchangers on the existing plant which transfer heat across the pinch or represent external heating below the pinch or external cooling above the pinch can be identified. These exchangers signify inefficiencies in the existing process, and alternative heat exchange matches can be investigated.

In this phase of the study, computer programmes which are now available to translate the stream data into composite curves can play a valuable part by permitting quick and efficient examinations of the effects of many different possible changes. The emphasis here is on the use of interactive programmes which enable the engineer to exercise his judgement and retain control of the design changes under consideration.

The second phase of the study would be completed by a preliminary screening of all the opportunities which have been identified whilst eliminating any which cannot meet economic and practical criteria. The resulting shortlist of projects is then subjected to a more detailed evaluation in phase three. This evaluation would include an assessment of the safety and operability of the proposed changes plus equipment definition and cost estimates to enable an expected payback period to be calculated.

Case Study of Process Integration Techniques

The potential benefits from the application of these techniques can be illustrated by summarising the results of a study which was carried out by specialist consultants on a modern grain whisky distillery with an annual energy bill of approximately £1.5 million. This distillery used a state-of-the-art process which had already achieved about 30 per cent energy saving compared with traditional plant.

The initial analysis of the process indicated that the external heating target was 82 per cent of the current consumption. However, by applying the plus/minus principle, two feasible process modifications were identified which reduced the external heating target to 72 per cent, corresponding to a potential saving of nearly £400,000 a year.

The final results of this study identified three separate projects, as shown in Figure 7.10. Project A achieved 3.5 per cent energy saving by a process modification which reduced a hot stream duty below the pinch. The stand alone payback period for this project was only six weeks and in fact the modification was implemented while the study was still in progress.

FIGURE 7.10: ENERGY SAVING PROJECTS IN RELATION TO CURRENT ENERGY USE AND EXTERNAL HEATING TARGET IN A DISTILLERY

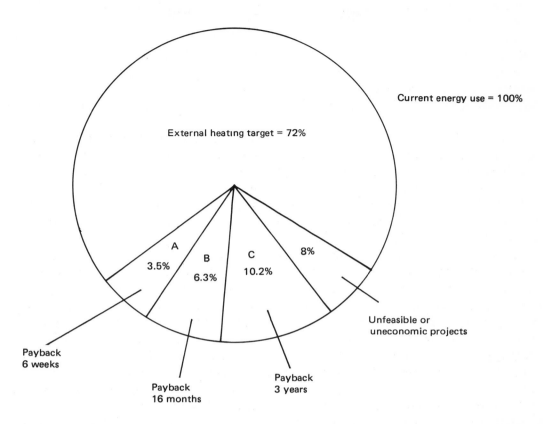

Project B utilised an opportunity to pump heat from a hot stream below the pinch into a cold stream above the pinch. This could be achieved by generating low pressure steam against the hot stream and raising this to an intermediate pressure in an eductor so that it could be condensed against the cold stream. When implemented, this project will save 6.3 per cent of the total energy bill at a payback of 16 months.

Project C provides a further saving of 10.2 per cent of current energy use at a payback of approximately three years. This project was identified by recognising that one of the heat exchangers on the existing plant was inappropriately placed because it transferred heat across the pinch. By eliminating this cross-pinch heat transfer both external heating and cooling duties can be reduced. At the same time this project offers additional benefits increasing plant capacity by eliminating a bottleneck.

The total energy saving which could be realised by implementing projects A, B and C is 20 per cent of the original use compared with the target of 26 per cent. In fact, other projects were identified which would have saved most of the extra six per cent, but these were rejected during the course of the study as either impractical or uneconomic.

A fourth project which was also evaluated as part of this study was the integration of a gas turbine cycle into the existing process. The turbine would generate the total distillery electricity base load providing an annual saving of about £120,000 a year at an expected payback of four years. This project will be re-examined after other projects providing shorter paybacks have been implemented.

The impact of this study on the total energy consumption of the distillery is shown schematically in Figure 7.11. The shape of this graph demonstrates features which would be seen in many cases where process integration technology is applied. Firstly, there is a learning curve where energy consumption has been gradually reduced by conventional means, and secondly, after pinch techniques have been applied and the energy targets set, there are step changes in energy consumption.

FIGURE 7.11: TYPICAL ENERGY CONSUMPTION VS TIME BEFORE AND AFTER APPLYING PINCH TECHNOLOGY

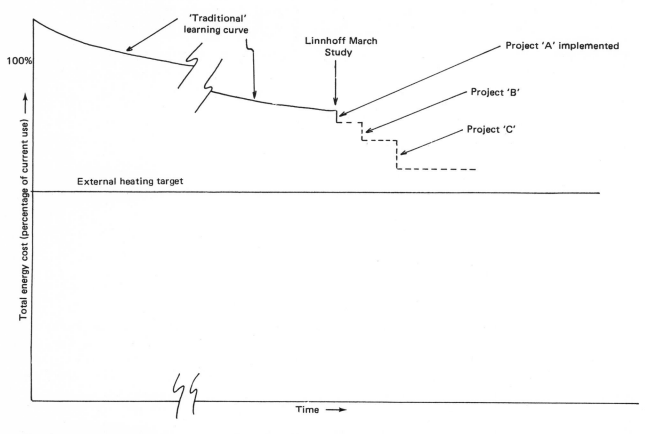

The basic principles of process integration are easily applied to optimise heat exchanger networks for simple continuous processes. They are now sufficiently developed for application to batch processes, integration of combined heat and power cycles, identification of fundamental energy saving process changes, optimisation of multiple utility systems and incorporation of plant safety and operability constraints.

In most applications the best way to exploit the full potential of process integration is to employ a consultant who is well versed in these techniques, is familiar with the latest developments, and has access to the appropriate computer software. Study costs are generally low and grants are often available.

8

NEW OPPORTUNITIES
FOR CUTTING ENERGY COSTS

David Wragg

The Energy Manager's Workbooks Volume 1 and 2 have covered in some detail a wide range of topics, each of which is an important part of the work of energy management. In this chapter we take a brief look at a number of areas which are growing in importance, partly as a result of technological developments and partly owing to the shifting economic environment in which we have to work.

Combined Heat and Power Generation

Any industrial or commercial enterprise has a need for two kinds of energy: heat and work. Both of these can be obtained from any fuel but the tendency in most enterprises is to buy them separately, as fuel for heating and electricity for power (work) and lighting. This has several advantages; it is flexible, it enables us as a nation to make good use of cheap coal, it exploits the benefits of large-scale power generation, and it does not require capital investment.

It is also inefficient. It means that industry is not making use of the available work in the fuel it buys and the electricity supply industry is throwing away heat in the generation of electricity.

Combined generation of heat and power (CHP) is not a new concept. At one time a great many industries—such as the paper industry, breweries and sugar refineries—generated their own power and utilised the associated heat. Some still do, but there has been a steady decline in the proportion of electricity needs which industry generates for itself. This has been for several reasons:

—the increasing reliability of the public supply

—the narrowing of the price difference between electricity and other fuels

—unfavourable tariffs for generation in parallel with the public supply.

Three recent changes have made combined heat and power more attractive. These are:

—the 1983 Energy Act, which obliged the Electricity Boards to issue tariff structures enabling organisations who generate their own electricity to sell electricity to the Boards. It also allowed them to use the National Grid system to transport the electricity, subject to safeguards being installed at the site between the generator and the Board's supply

—the availability of small packaged engine-driven CHP units, such as those set out in Table 8.1

—grants for investigating the feasibility of CHP systems.

Factors Affecting the Viability of CHP

Efficiency: The improvement in efficiency using the CHP system, as against the separate provision of heat and the public supply of electricity, needs to be sufficient to balance the higher cost of fuel and the capital cost of the system.

TABLE 8.1: OUTPUT OF SOME SMALL PACKAGED ENGINE-DRIVEN CHP UNITS

	Thermal Output (kW)	Electrical Output (kW)
Fiat Totem	28	15
Serck Heat Transfer	28 and 86	15 and 36
Holec	51-256	20-150
Applied Energy Systems	59-277	24-132

Heat to Power Ratio: Separate provision of heat and power is very flexible. Most CHP systems have a fixed optimum ratio of heat to power and a narrow range of ratios over which they can work.

Characteristics of the Heat Load: The heat made available must be suitable. If the heat output of the CHP system is mainly low grade heat, as is the case with small engine-driven systems, there must be a suitable application for it.

Fuel Choice: The choice of drive system and the choice of fuel are often interlinked. Engine-driven systems are dependent mostly on gas or distillate oil fuels. Coal is more usual with steam drives.

Tariff Structure: This will depend on the type of electricity requirements, i.e. whether it is to be exported (sold to the public supply), or operation is parallel to the public supply, or standby is needed.

Annual Utilisation: A high annual utilisation is usually required to offset the high capital costs.

Maintenance: Some of the more complex systems may require a high level of maintenance.

The Efficiency of CHP Systems

The relative efficiencies of different CHP systems are shown in Figure 8.2. The overall efficiency of steam power stations in the public supply is about 28 per cent for conversion of fossil fuel. A typical modern package boiler provides heat at about 80 per cent. This is one reason why electricity remains one of the more expensive common fuels.

The importance of the heat to power ratio for the selection of the system is clear from Figure 8.2. The outputs of some representative engine-driven systems has already been given in Table 8.1.

CHP Feasibility Study Applied to College Buildings

In the Energy Manager's Workbook Volume 1, college buildings were used as an example of a practical exercise in energy auditing (Chapter 9). This gave information on the electricity and fuel consumption month by month, which could be used to investigate whether or not a combined heat and power system would be viable for the college.

Table 8.3 gives the monthly fuel oil consumption in litres and Table 8.4 gives the monthly electricity consumption and maximum demand with the updated costs.

The college has since taken advantage of the tariff allowing cheaper night units. The total updated cost should therefore be £10,812 assuming 20 per cent of the total units are used at night.

FIGURE 8.2: SANKEY DIAGRAMS SHOWING THE EFFICIENCIES OF SOME POWER GENERATING SYSTEMS

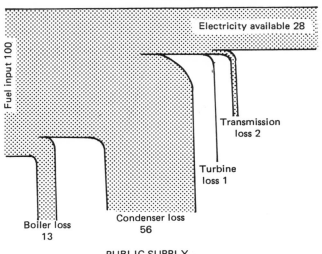

PUBLIC SUPPLY

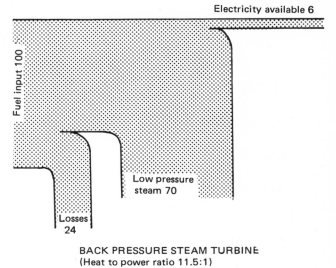

BACK PRESSURE STEAM TURBINE
(Heat to power ratio 11.5:1)

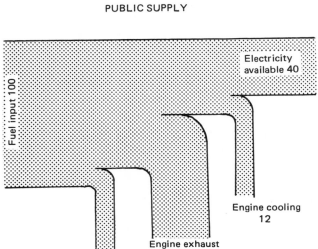

DIESEL GENERATOR WITH ENGINE COOLING AND
EXHAUST HEAT RECOVERY
(Heat to power ratio 1.15:1)

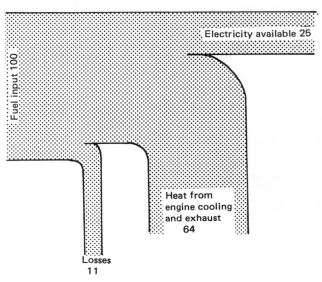

GAS ENGINE GENERATOR WITH ENGINE
COOLING AND EXHAUST HEAT RECOVERY
(Heat to power ratio 2.5:1)

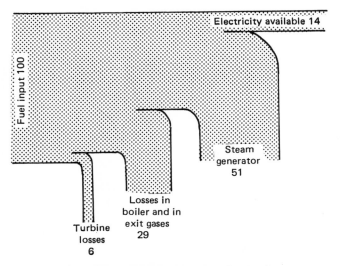

GAS TURBINE WITH WASTE-HEAT BOILER
(Heat to power ratio 3.6:1)

TABLE 8.3: MONTHLY FUEL OIL CONSUMPTION

	Litres	£
January	36,000	6,480
February	31,000	5,580
March	28,000	5,040
April	20,000	3,600
May	15,000	2,700
June	4,000	720
July	4,000	720
August	4,000	720
September	5,000	900
October	22,000	3,960
November	30,000	5,400
December	29,000	5,220
Total	228,000	41,040

TABLE 8.4: MONTHLY ELECTRICITY CONSUMPTION

	kWh	Max kVA	Monthly Cost £
January	33,460	88	1,569.91
February	27,820	84	1,352.03
March	31,340	92	1,337.18
April	17,510	78	693.86
May	17,970	84	710.53
June	26,420	78	1,016.46
July	18,950	70	746.00
August	12,660	58	518.28
September	18,380	62	725.37
October	20,420	72	799.23
November	24,210	76	1,054.24
December	27,250	86	1,340.62
Total	276,390		11,863.71

Average Cost = 4.3p per kWh

Heat/Power Ratio

The first step in the assessment is to examine the heat/power ratio. To do this, we must express the heat and power in consistent units. In this case, we will adopt kilowatt hours, as electricity is already expressed in these units. We must also make an allowance for the efficiency of the boiler in converting fuel to heat, for which we have assumed 65 per cent. This is shown in Table 8.5.

TABLE 8.5: RATIO OF HEAT GENERATED IN COMPARISON TO ELECTRICITY USED

	Electricity kWh	Heat kWh	Heat/Power Ratio
January	33,460	247,320	7.4
February	27,820	212,970	7.6
March	31,340	192,360	6.1
April	17,510	137,400	7.8
May	17,970	103,050	5.7
June	26,420	27,480	1.0
July	18,950	27,480	1.4
August	12,660	27,480	2.2
September	18,380	34,350	1.9
October	20,420	151,140	7.4
November	24,210	206,100	8.5
December	27,250	199,230	7.3

The fuel currently in use in the college is diesel oil. It would be convenient if we could use the same fuel for electricity generation, as this reduces the need for further storage or for buying stocks of a different fuel. With full heat recovery from the engine cooling system and the exhaust gases, the diesel generator provides heat and power at a ratio close to 1:1. That is, for each kilowatt of electricity generated there will be one kilowatt of heat available. From Table 8.5 we can see that the heat/power ratio is always greater than this. To proceed further we must now evaluate the cost of generating electricity.

Cost of Generating Electricity

As the loading of the system is reduced the efficiency of a diesel engine generator falls, with a consequent rise in the cost of generation. When heat is recovered the savings incurred can be offset against the cost of electricity. This is illustrated in Figure 8.6 which shows the cost of electricity generated over a range of engine loadings where:

a) there is no heat recovery

b) there is 50 per cent utilisation of the recoverable heat

c) there is 100 per cent utilisation of the recoverable heat.

Also compared in the diagram are the costs of buying electricity from and selling it to the public supply, through the East Midlands Electricity Board.

From this diagram we can draw a number of conclusions:

—the cost of private generation between 0030 hrs and 0730 hrs is always greater than the price to buy electricity from or sell it to the public supply. So, generating your own electricity at night is not cost effective in this case

FIGURE 8.6: COST OF GENERATING ELECTRICITY USING A DIESEL ENGINE COMPARED TO PUBLIC SUPPLY TARIFFS

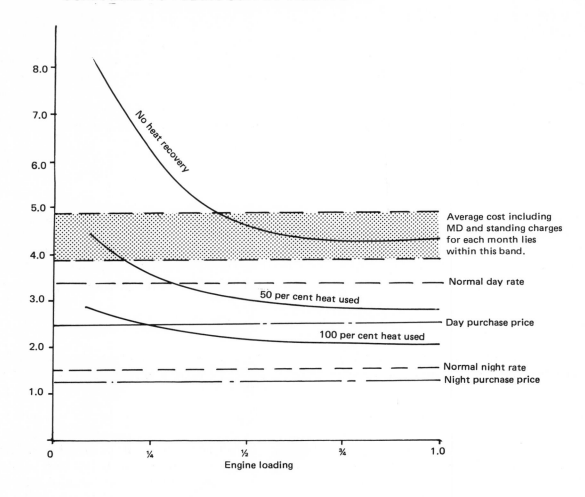

FIGURE 8.7: ELECTRICITY DEMAND PROFILE FOR A SEPTEMBER WEEKDAY

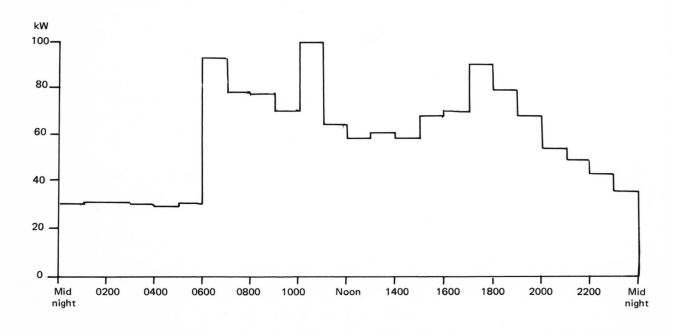

—operating the generator without heat recovery will only be cost effective for short periods of maximum demand in December, January and February, for about one to one-and-a-half hours a day

—increased generation costs will occur at less than 50 per cent load, and below 25 per cent load this increase is severe

—in an extreme case, if all of the recoverable heat were used in the college and any surplus electricity sold to the public supply, the generator would make a profit of about 0.5 pence per kWh sold to the board.

In this case the extra cost of metering and tariff charges would render selling electricity economically unattractive.

Electricity Demand Profile

It is important when deciding on the size of generator and loading to know how the electricity demand varies during the day. Figure 8.7 shows the demand profile measured for a September weekday, which is reasonably typical. This shows that the average night load (between 0030 hrs and 0730 hrs) is 39 kW and the average daytime load (between 0730 hrs and 0030 hrs) is 66 kW. The total consumption on night tariff is 20 per cent: for our assessment this percentage is assumed throughout.

Heating Demand

It is more difficult to measure the heat demand profile. From the monthly fuel oil consumption we can deduce that the average summer heat demand is 50 kW and the average winter heat demand is 304 kW.

Preliminary Assessment of the Case for Private Generation

From the above it is clear that the most viable case for self generation will occur when all the recoverable heat from the generator is used. If a generator rated at 40 kW (electrical) were selected it would produce 40 kW of electricity and approximately 40 kW of hot water at full load. This could be used as a base-load throughout the year, as the summer heating load could adequately absorb all the available heat. Thus, electricity would be generated at minimum cost, i.e. two pence per kWh.

A 40 kW installation cannot meet the total daytime electricity demand, as all monthly maximum demand figures are above 40 kW. However, from the calculated daily average demand for electricity an estimate can be made of the proportion of demand that could be met. It appears that a 40 kW installation would meet about 90 per cent of demand.

Average cost of purchased electricity	= 4.3 pence per kWh
Average cost of generated electricity with full heat recovery	= 2.0 pence per kWh
Saving	2.3 pence per kWh

So the saving through private generation is about 53 per cent of current costs of buying from the public supply. The total cost saving would be £5,730 per annum. The generator would cost about £13,000, so on a preliminary assessment the payback period is a little more than two years and there would be a case for private generation.

A More Detailed Assessment

It is now necessary to take into account the detailed tariff structure, including the fixed and maximum demand components which will reduce the level of savings, and to assess other costs incurred with the installation of a generator.

It is important to examine the electricity charges in each month. For example, in January:

Total electricity requirements = 33,640 kWh
Night-time requirements   =   6,692 kWh
Daytime requirements   = 26,768 kWh

Total electricity generated (for 17 hours a day, 31 days in the month)
= 40 x 17 x 31 = 21,080 kWh

This is less than total requirements so that all of the electricity can be used. The maximum demand charge for the month will be based on the current maximum demand, less the 40 kW we generate i.e. 48 kVA. The daily demand profile shows a maximum demand peak occurring before 0730 hrs. So to reduce maximum demand charges it would certainly be worth operating the generator for an extra hour to encompass this peak during December, January and February. The total annual costs are shown in Table 8.8.

TABLE 8.8: ANNUAL COST OF ELECTRICITY TO THE COLLEGE USING PRIVATE GENERATION, WITH A 40 KW GENERATOR

		£
Generated Electricity	203,422 kWh @ 2.0 pence	4,068.44
Purchased Electricity	Night	851.28
	Day	458.85
	Fuel Adjustment	168.54
	Maximum Demand	599.34
Monthly Charge	12 x £70	840.00
Availability Charge	53p x 100 kVA x 12	636.00
Fixed Charge	5.1p x 100 kVA x 12	61.20
	Total Cost	£7,683.65

Annual Savings = £10,812.32 – £7,683.65 = £3,128.67

This immediately lengthens the payback to over four years. The additions to the capital cost which we need to consider are:

—switchgear
—installation
—modifications to buildings etc
—heat storage to maximise use of recovered heat.

And we also need to include costs of additional maintenance and downtime.

We can see that generating electricity, even with full heat recovery, could not be justified in this case. However, it clearly shows that to maximise the viability of the case all the recoverable heat from the engine must be used. Also, the availability of natural gas would reduce the running costs considerably.

Heat Pumps

The heat pump is not a new concept. In some countries heat pumps have been used for space heating since the 1930s. The recent interest in heat pumps has followed the growing awareness of various types of heat recovery, and the fact that it is now practical to consider forms of drive system other than the electric motor.

Among the newer applications for heat pumps are:

—combined process heating and cooling
—drying and dehumidification
—waste heat recovery.

What is a Heat Pump?

A heat pump is a device which takes heat from a low temperature source such as air or water and upgrades it so that it is available for use at a higher temperature. The most common type of heat pump is the vapour compression type which has five main components:

—evaporator
—compressor
—expansion device
—condenser
—working fluid.

Figure 8.9 shows how the compression cycle heat pump works. Beginning at point A in the diagram, the working fluid is drawn into the compressor. Compression causes the temperature of the working fluid to rise. The heat is then extracted from the working fluid as it passes through a heat exchanger, between points B and C. Passing the vapour through the expansion device causes its temperature to fall. The cool fluid can now be used to take up heat from a heat exchanger as it passes from C to A. The cycle then recommences. The heat pump is thus a means of transferring heat from one heat exchanger to another and at the same time raising its temperature.

FIGURE 8.9: THE VAPOUR COMPRESSION HEAT PUMP CYCLE

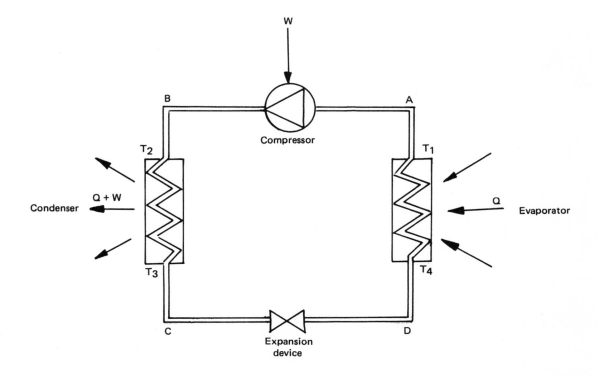

The working fluid is usually chosen so that it is liquid in part of the cycle and vapour in the rest. In the heat exchanger where it gives up its heat it changes from vapour to liquid and the heat exchanger is called the condenser; in the heat exchanger from which it extracts heat it reverts from liquid to vapour and the heat exchanger is called the evaporator.

Efficiency

The efficiency of a heat pump is described in terms of its Coefficient of Performance (COP). The ideal COP can be derived from thermodynamics and is expressed as:

$$COP = \frac{Q + W}{W} = \frac{T_2}{T_2 - T_1}$$

where Q is the quantity of heat moved
W is the power absorbed by the compressor
T_2 is the temperature of the evaporator
T_1 is the temperature of the condenser

(temperatures must be expressed as absolute temperatures to calculate the COP).

For example, if the evaporator temperature is 10°C (283 K) and the condenser temperature is 40°C (313 K) the ideal COP is:

$$\frac{313}{313 - 283} = 10.4 = \frac{Q + W}{W}$$

Therefore, for each unit of power (W) absorbed by the compressor the total heat delivered at the condenser (Q + W) is 10.4

The COP achieved in practice will be something like one third the ideal figure, due to compressor inefficiencies, non-reversible expansion, losses in the pipework, and temperature losses in the heat exchangers.

The coefficient of performance is crucial to the economics of a heat pump installation since the highest possible heat output will be sought for the minimum energy input to the compressor. When contemplating a heat pump installation it is important to bear in mind that:

—as the temperature lift (the difference between the condenser and evaporator temperatures) increases, the COP falls. A heat pump will operate more efficiently, and hence could possibly be made more economically attractive, by using some simple heat exchange to reduce the temperature lift (see Figures 8.11 and 8.12)

—the term coefficient of performance strictly speaking refers to the heat pump cycle, but it is occasionally used to describe the overall performance of an installation. This can give rise to confusion when describing engine-driven systems. When approaching a manufacturer ensure that it is understood what is meant by the figures quoted.

Choice of Drive System

There are several different means available to drive the compressor, including:

—electric motor
—gas engine
—diesel engine
—gas turbine
—steam turbine.

The choice of drive system is also a critical factor in determining the operating economics since it affects the cost of fuel required to run the heat pump, and the capital cost of the equipment. Engines and turbines are more expensive than electric motors, but electricity is an expensive form of energy.

Invariably, the economic case for an engine-driven or gas turbine driven heat pump will require good use to be made of the reject heat from the drive system.

As an example, the overall heat balance of a heat pump system (whose COP is assumed to be four) driven by a gas engine would be:

Input
    Fuel to engine   =   1 unit

Output
    0.29 units available at the shaft x 4   =   1.16
    0.21 units engine cooling (all recovered)   =   0.21
    0.40 units engine exhaust (60 per cent recovered) =   0.24
    0.10 units losses not recoverable   =   0.0
                                            ——————
                      Total Output   =   1.61 units

Figure 8.10 shows a fuel cost chart which enables a ready comparison to be made between a heat pump and conventional heating methods. In calculating the coefficient of performance referred to in the diagram it is important to base it on the total useful output of heat and the total energy input. That is, it must allow for any heat recovered from engines, latent heat absorbed from drying and dehumidification systems and power used for pumps or fans, etc.

Capital Costs

In this context it is only possible to give an order of cost so that a rough comparison can be made between the heat pump and the more conventional forms of providing heat.

Taking a natural gas-fired hot water boiler with a rating of 200 kW as unity, the following order of costs apply:

Natural gas fired boiler	1.0
Light fuel oil boiler	1.1
Solid fuel boiler	1.9
Electric heat pump	3.6
Gas engine heat pump	4.7

Stages in the Appraisal of a Heat Pump Installation

The most direct way to assess the feasibility of a heat pump installation is first to assign a value to the heat obtainable, and then to compare this with the cost of running a heat pump to produce it. Assuming a source of waste heat has already been identified, the steps involved are:

1. Find out if the amount of waste heat can be reduced, by improving the efficiency of the process from which the heat is produced, or by use of simple heat recovery (i.e. without heat pumps).

2. Calculate the amount of heat available and determine its temperature. The variation in the load should be determined in case it is necessary to use some heat storage. This will be important in the final selection of the heat pump size and system management.

FIGURE 8.10: COMPARISON OF FUEL COSTS USING A HEAT PUMP TO FUEL COSTS USING CONVENTIONAL METHODS

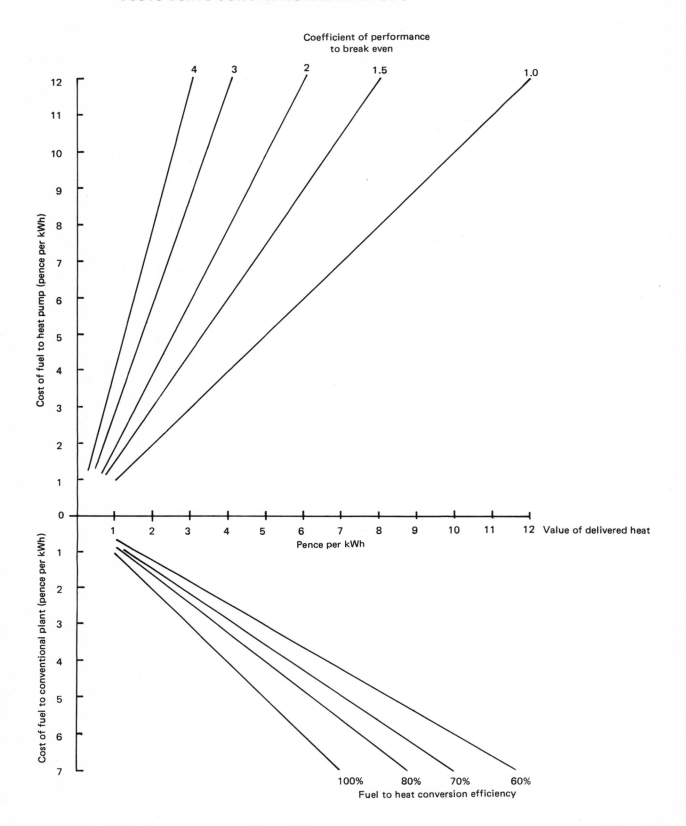

3. Estimate the expected coefficient of performance for the temperature at which the output heat can be used.

4. Calculate the quantity of heat available from the heat pump.

5. Determine the value of this heat in terms of fuel replacement costs, not forgetting to allow for the efficiency of boilers etc used to convert fuels.

6. Calculate the size of drive system required and its running costs.

This information should be sufficient to enable a preliminary evaluation of the case and an estimate of the capital cost which can be justified for a heat pump installation.

Optimising Heat Pumps

At this stage it is pertinent to discuss COP in more detail, in order to see how we can optimise the value for any given situation. This will be done in the form of a few examples, which are illustrated in Figures 8.11-8.13.

Base Load Heating: This is a fairly common way of ensuring that the heat pump operates at, or near, maximum load. The dynamics of the connected load need to be considered here.

The connected load may only exceed 75 per cent of the full load for a small part of the time in which case it may pay to size the heat pump for 75 per cent of the load, and install alternative heating plant to supplement when necessary. This will alleviate the effect of part load operation, which tends to reduce the overall COP, and will obviously affect the economic case for installation.

Simple Heat Exchange: Another means of raising the COP of a system is by incorporating a simple conventional heat exchanger in the heat recovery system. The temperature lift remaining for the heat pump will be minimised and the COP maximised.

Cascade Heat Pumps: If it is necessary to heat or cool a stream through a large temperature range, it may be worth investigating the use of a cascade system which integrates two or more heat pumps. Each heat pump will have a higher COP (and lower temperature lift) than a single unit doing the same job, and the combination will have an overall COP higher than the single unit.

Choice of Drive

The nature of the sink (where the heat is delivered) and source may well have a bearing on the type of drive chosen. For example, if exhaust air from a dryer has to be lifted from 20°C to 70°C an electrically driven heat pump will only give a COP of about two to three. A gas engine driven heat pump can operate at a COP of about five if the heat pump lifts the temperature to 50°C, the final 20°C being met by the engine exhaust.

Summary

Heat pumps are a definite advantage in certain applications but they require considerable capital investment and each case must be considered carefully. Expert advice must be sought, certainly at the design stage and probably at the assessment stage, to ensure that all avenues have been investigated.

Although this presentation is necessarily brief, it demonstrates that the options for heat recovery using heat pump techniques are numerous and that knowledge of the characteristics of the connected load is essential if the system selected is to give the required output at the optimum COP and hence the lowest running cost.

FIGURE 8.11: HEAT PUMP SYSTEM USING A STANDBY BOILER TO TOP UP
BASE LOAD

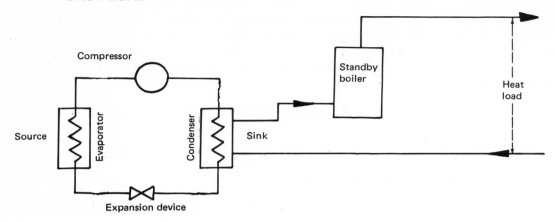

FIGURE 8.12: HEAT PUMP SYSTEM USING HEAT EXCHANGER TO RAISE TEMPERATURE
OF SINK

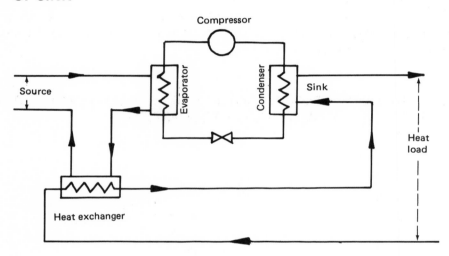

FIGURE 8.13: CASCADE HEAT PUMP SYSTEM

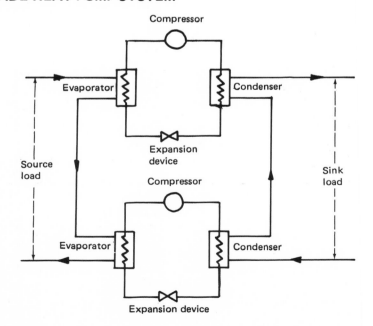

Induction Motor Controllers

A large proportion of electricity consumption—up to 30 per cent—is used in electric motors. Most of these are induction motors, the most common type of which is the squirrel-cage induction motor. This type is popular because it is extremely robust, is cheap to make and has virtually no maintenance costs. However, the features of the design which give us these advantages also pose certain problems:

—they are constant speed motors, the speed being determined by the frequency of the electricity supply

—they are efficient when fully loaded, but become less efficient with low loads

—they have poor power factors when lightly loaded

—they have high starting torques and it is not uncommon for designers of plant incorporating motors to oversize the motor.

When we have no means of exercising control over the motor in use, these drawbacks lead to inefficient running. This is shown in Figure 8.14.

The fall in power factor with motor load is utilised by a relatively new energy saving device to reduce partial load losses in induction motors. These controllers can be easily retro-fitted between the existing motor starter, and motor

FIGURE 8.14: VARIATION OF POWER FACTOR AND EFFICIENCY WITH LOAD USING A TYPICAL 15 KW, 3 PHASE, 4 POLE MOTOR

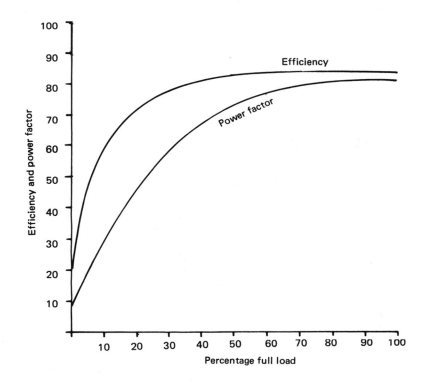

Newer methods incorporate electronic techniques. When first introduced as 'power factor controllers', they used analogue electronic circuits and required considerable on-site installation and matching to individual motor characteristics. More recently, electronic digital techniques combined with fast action switching have made it possible to control motors by regulating the frequency of the electricity supply. The advantage of this development is that it can be applied to motors already in use.

Figure 8.15 shows the levels of savings which can be achieved using an induction motor control. Table 8.16 shows some typical results taken from a factory survey carried out by NIFES.

Some care needs to be exercised in assessing the savings for a given motor, as different savings may be made with different types of motor of equal output rating, performing equal duties. This is for several reasons:

— the difference in normal full load and part load efficiencies of the motor

— the distribution of losses in the motor, in particular the ratio of iron loss to copper loss

— the sensitivity of the motor losses to harmonics in the supply.

Other benefits attributable to this controller is its ability to reduce power surge on motor starting and consequent burn outs.

FIGURE 8.15: POTENTIAL SAVINGS USING DEVICES TO IMPROVE THE EFFICIENCY OF ELECTRIC MOTORS

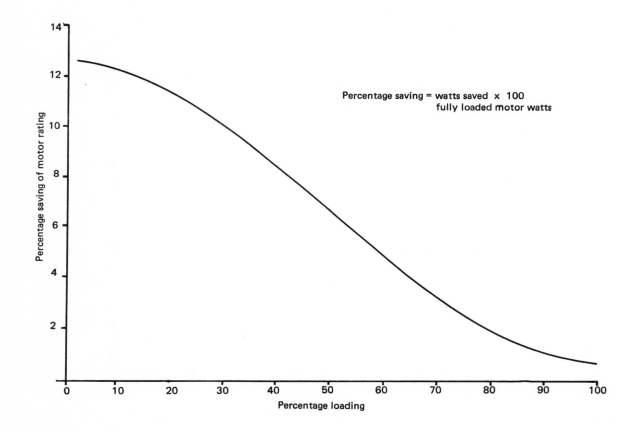

$$\text{Percentage saving} = \frac{\text{watts saved} \times 100}{\text{fully loaded motor watts}}$$

TABLE 8.16: POSSIBLE SAVINGS IN A NUMBER OF MOTORS FOUND IN A FACTORY SURVEY

Motor	Rating	Mean Load	Expected Savings		Capital Cost	Simple Payback
	kW	kW	per cent	£	£	yrs
Screw compressor	110.0	110.0	Nil	Nil	—	—
Boiler fan	30.0	20.7	4.0	284	650	2.3
Boiler fan	30.0	20.0	3.9	280	650	2.3
A/C fan	45.0	26.5	5.6	751	912	1.2
Water pump	18.5	11.1	5.5	201	460	2.3
Air heating unit 1	11.0	7.4	4.2	79	380	4.8
Air heating unit 2	11.0	7.4	4.2	79	380	4.8
Air heating unit 3	37.0	29.0	2.5	159	780	4.9
Air heating unit 4	37.0	29.0	2.5	159	780	4.9
Air heating unit 5	22.0	13.2	5.5	240	560	2.3
Water pump	18.5	8.9	8.0	290	460	1.6
Water pump	18.5	8.3	8.1	295	460	1.6

Controlling Ventilation Rates by Monitoring Carbon Dioxide

Many buildings lose large amounts of heat because there is no means of controlling ventilation rates or relating the degree of ventilation to requirements. This is a common problem in buildings where the number of people who may be in the building at one time is subject to great variety.

Since all the ventilation air must be heated to the space temperature, there is clearly scope for energy savings by controlling the rate of ventilation to correspond to the requirements of the occupancy level. In buildings such as cinemas, theatres, department stores, in fact most public buildings, the degree of occupancy can fluctuate throughout the day.

Other common problems with ventilation are:

—the number of occupants is less than that for which the system is designed

—infiltration of air is not generally allowed for in the design of the building

—air quality does not deteriorate instantaneously after initial occupancy.

The idea of controlling the ventilation rate by monitoring the levels of carbon dioxide (CO_2) in the air inside the building has attracted considerable interest. CO_2 concentrations in the space can be related to the ventilation rate per person. The CO_2 concentrations in the outside air are fairly constant and for a given activity rate the production of CO_2 by people is known. It follows therefore that by measuring CO_2 concentrations the number of occupants can be monitored and the ventilation rate adjusted accordingly. In fact the device controls the CO_2 to a constant level.

One of these devices has been tested in a department store where over 90 per cent of the heating is provided by the ventilation system when operating at design conditions. Claims have been made for this case of a payback period of less than two years.

The following points are worthy of note:

—internal heat gains tend to lower the savings since the ventilation system may have to operate in the 'free cooling' mode (i.e. full fresh air) for a proportion of the time

—there may be spin-offs such as reduction in cooling load for air conditioned systems when the enthalpy (heat content) of the external air is greater than the enthalpy of the internal air

—adjustments may have to be made to the design CO_2 level to allow for smoking areas. There does not yet appear to be much information relating to this

—priority must be given to 'free cooling' requirements of an air conditioning plant otherwise the cooling load will devour any savings

—the CO_2 monitoring and control equipment must be fail-safe, i.e. the fresh air damper must go to the full open position.

An electronic control unit installed in a cinema auditorium to measure and control the CO_2 level, is shown in Figure 8.17.

FIGURE 8.17: VENTILATION CONTROL BY THE MEASUREMENT OF CARBON DIOXIDE LEVELS IN A CINEMA AUDITORIUM

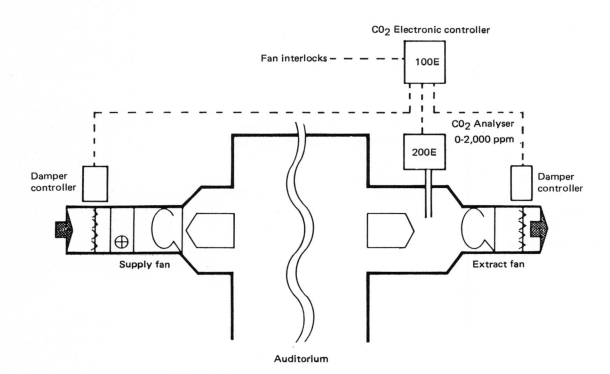

HEAT MANAGEMENT SERVICES

Alan Tweedale

The idea of heat management was developed almost by accident in the early 1930s, when the General Hospital in Lille, France, was on the point of closing because of an industrial dispute. It was at this stage that the local heating contractor was called in to operate the boiler plant. Today that company—Compagnie Generale de Chauffe—is the largest heat management company, and has a current turnover in excess of £215 million. Despite the acceptance of heat management on the Continent, however, it was not until the middle 1960s that this concept was introduced into this country, when the National Coal Board formed Associated Heat Services in conjunction with Compagnie Generale de Chauffe.

Heat management means exactly what it says—the day-to-day control of a client's existing boiler plant and heating system. In parallel with heat management is a virtually identical activity which deals with the supply of heat, steam and electricity as a utilities supply service. In this latter case the equipment used to produce the heat, steam and electricity is both operated and owned by the heat management company.

The primary objective of heat management is to ensure that the fuel resources, labour and materials consumed are kept to an absolute minimum, whilst still maintaining a reliable and continuous utilities supply of heat, steam or electricity. In this way the useful energy will be available at the point of usage at the lowest possible cost.

Principal Services Provided

The principal services provided are the supply of heat or steam to customers' plant or premises and, increasingly, the provision of air-conditioning. A heat service company can attend to the operation and maintenance of boilerhouses, boiler plant and heat distribution systems, supply the requisite fuel to its customers' sites, or generate heat or steam using the customers' own boilers and ancillary plant. Operatives make regular service visits and offer a prompt repair service to cater for breakdowns. In some cases, control is maintained through on-site monitoring equipment owned and supplied by the heat service company and linked by telephone lines to their operational centres which are manned 24 hours a day.

The company will also undertake to replace certain items of plant and equipment under a replacement guarantee scheme (referred to below).

For those customers who do not require a full heat or steam supply service, a heat management company can usually provide a maintenance-only service, which involves regular visits as well as prompt repairs in the event of breakdowns. Additionally, companies usually offer a design and construction service in relation to boilerhouses, boiler plant and heat distribution systems, often by co-ordinating the work of sub-contractors.

Potential Advantages

Heat management companies provide their clients with clear financial benefits through the bulk purchase of fuel and sundry materials and the availability of the skilled engineers and boiler operators who run and control the boiler plant and heating systems. In many areas, one boiler operator can look after up to a dozen small installations, thus providing the customer with great savings compared with the cost of employing full-time personnel. Furthermore, by the use of up-to-date monitoring techniques, the company is able to ensure the continuous efficiency of boiler operations, with resultant economies for the customer.

Customers fall into three main categories:

—industrial customers, who require supplies of heat for factories and offices, and steam for industrial processes

—commercial customers, who are supplied with heat for offices

—local authorities and housing associations, to whom heat is supplied for offices, district heating schemes, schools etc.

Range of Services Offered

The various items usually included in the standard service of a heat management company will now be examined.

Fuel

All fuel used to generate heat, steam and electricity is included within the charges. The fuel is specified, checked on delivery for quality and quantity, and paid for by the heat management company. It is usually purchased under carefully negotiated terms which fully reflect the bulk purchasing power of the company, the benefits of which are passed on to the customer. Great care is usually taken to ensure that strategic supplies of fuel are maintained, and frequently various sources and supplies of fuel are used to service one contract, thus reducing the risk of interruptions resulting from industrial disputes in the fuel and transport industries. Because of the large number of plants under operation, this can be achieved without detracting from substantial fuel rebates.

Whether the fuel chosen is oil, gas or coal, it will be constantly monitored by the heat management company and any conversion from one fuel to another which gives a cost advantage can be implemented. The heat management company will ensure that the payback period on conversion capital is not of longer duration than the fuel supply agreement.

The major heat management companies are constantly seeking ways of obtaining lower priced fuels. The use of imported and blended fuels—which may be slightly more inconsistent than UK produced fuels—often has an appreciable price advantage, and the resultant lower costs of heat and steam are passed on to the client.

Fluidised bed plant, which burns refuse-derived pelletised fuel and requires a high degree of skill and expertise in plant operation, is an ideal application for heat management.

Labour

The labour needed to look after the day-to-day requirements of the plant is usually provided by the heat management company on a directly employed basis. The practice of sub-contracting both labour and maintenance work does not appear to be successful, and if the heat management company does not have full control of all staff, then efficiency and reliability inevitably suffer. Larger coal-fired plants require full-time operation but even steam raising coal-fired plant up to 6 MW may be attended on a part-time basis.

The plant operators are usually mobile, with communication systems in their vehicles for instant call back to any plant. At all times the plants are fully monitored to raise the alarm if any mechanical or electrical malfunction occurs; the calls usually go to a control centre which is always working twenty-four hours a day. Operator duties include de-ashing the boiler daily or hourly (depending upon size), checking boiler efficiency, cleaning, carrying out routine water treatment tests and dosing the boiler etc.

Maintenance of the boiler plant and heating system is carried out by the heat management company's own staff, using carefully planned and well proven maintenance schedules. Plant repairs—whenever or however they occur—are also included in the service.

A reputable heat management company will have its own training facilities for all levels of industrial staff, to ensure that they are conversant with company policy and fully understand combustion principles and the needs of modern plant.

Securing and retaining reliable high standard plant operators has never been easy for individual organisations, but a structured career for all levels of boiler plant and service staff exists within a heat management company. This provides the incentive for greater efficiency of plant operation, and the benefits of better control and optimum performance are part of the service offered to the customer.

The change from oil- and gas-fired boiler plant to coal-fired plant is likely to be slow, but progressive and almost inevitable. So will be the gradual use of waste materials such as pelletised industrial and municipal refuse, shredded waste, coal slurry etc. Use of these lower grade materials will require a higher degree of specialist operational staff and skilled labour, often on a full-time basis. These changes increase the justification for the use of heat service companies—indeed it is unlikely that many companies will proceed with these types of projects without a specialist operating organisation.

Whilst incineration plant, complete with heat recovery equipment, is almost standard on the Continent and has been for many years, there are few applications in the UK. A notable exception is the Nottingham District Heating Scheme's incineration plant and coal-fired back-up station operated by Associated Heat Services, which produces both heat and electricity. Encouraged by government grants, incineration plants are now being designed for use in UK projects, complete with heat recovery systems. The larger and more sophisticated plants require a highly reliable and economical supply of raw materials, thus again calling for the skills of heat management companies.

Consumable Items

All cost items such as water treatment, maintenance equipment, i.e. drive belts, greases, bulbs, paint etc, and repair materials are included in the charges. Replacement parts are referred to separately in the section dealing with replacement guarantee. The cost of providing the 24 hour monitoring service, public liability and engineering insurance cover, ash removal etc is usually included in the consumable items section, along with direct overhead costs of telephones, rent or rates on the boiler plant, sickness benefit, national insurance etc.

Budgetary Controls

One of the most important aspects in the running of all businesses is the close control of sales, product revenue and production costs. A heat management company operates in precisely the same way. The costs of fuel, labour and consumable items are compared, actual against phased budget, on a weekly, monthly and cumulative basis year by year. The historic records built up become invaluable in setting new targets. Each member of the operational team will be held to account for deviations from the budget and this accountability, usually on a weekly basis, is one of the major factors which determine the efficient performance of the heat management company.

Replacement Guarantees

The heat management company offers a special facility whereby a replacement guarantee is incorporated into heat service contracts; under this, they are responsible during the term of such contracts for the repair and replacement of certain items of boiler plant and equipment, as and when necessary. In these circumstances, a premium, calculated on a formula basis, is built into the contract price to cover the cost of implementing the guarantee.

The customer who opts for this guarantee benefits from an element of insurance against plant breakdown and the cost of replacement, especially as the heat management company is better qualified to judge the extent of any such costs. On the other hand, if the plant is carefully maintained and operated by its staff, the company can benefit from the premium in the contract price to the extent that it is not required to replace plant and equipment in full. A strong incentive exists therefore to prolong the plant by skilled and planned maintenance techniques.

The Energy Capsule

A novel concept of utilities supply for heat or steam which was developed by Associated Heat Services is the Energy Capsule (See Figure 9.1). This is a fully containerised boiler plant, usually fired by coal or low grade fuel, complete with overhead fuel silo, feedwater tanks, free standing chimney, interconnecting pipework, all electrical wiring, de-mineralisation plant, automatic blow-down vessel, fuel handling and fuel feed mechanism etc.

FIGURE 9.1: ENERGY CAPSULE INSTALLED AT WAKEFIELD COUNTY GENERAL HOSPITAL

The smaller units (up to 850 kW) are fitted into a capsule container of the same size as vehicle container units and are transported to site and off-loaded in one piece, with the chimney installed literally minutes later. The larger units, often fitted with fluidised beds of up to 5 MW capacity, are transported to site using a convoy of vehicles, and can, in theory, be operated within 48 hours of arrival on site.

The sole responsibility for the design, finance, complete operation, maintenance and replacement of plant lies with the heat service company, thus relieving the client user of all involvement. Any shortfall in efficiency has no effect on the metered heat or steam charges to the client. In many cases the client would retain his existing oil or gas fired plant for use as standby or for peak loading etc, thus obtaining true multi-fuel flexibility to give protection against strikes, transport disputes, supply restrictions, and so on.

Multi-boiler installations usually have a complete standby unit and in a three boiler installation, for example, it is probable that one boiler can carry up to 80 per cent of the annual load. Thus, a single energy capsule correctly sized can carry the major load of most industrial requirements, although of course additional space in an adjacent car park or works yard is required.

It is possible that the concept of factory produced energy capsules may have far-reaching effects upon the way in which future boilerhouses are constructed and plant installed. It is indisputable that the manufacture of skid mounted boilers, fuel silos, feed tanks, water treatment plant, chimneys, interconnecting pipework and electrics can be produced more neatly, cheaply and quickly in a controlled factory environment. Economies of scale plus the eventual higher recovery value of the plant in the event of factory closure are further advantages to be gained by this system.

Energy capsules are usually linked to a fully manned control centre which constantly monitors all operating parameters such as fuel stock, fuel feed rate, heat or steam output, make-up water feed, stack carbon dioxide and stack exit temperature, temperature or steam pressure and power failure. Alternatively, the plant can be remotely operated. This has now developed to such a degree that the fluidised bed energy capsules are fully automatic and monitored constantly, without the need for daily attendance by boiler plant operational personnel.

Methods of Charging

The most common methods of charging for the provision of heat or steam are:

- —fixed annual charge
- —two-part tariffs
- —metered charges.

These will now be discussed in turn.

Fixed Annual Charge

This is the simplest method of charging, but it is only applicable when it is possible to accurately calculate the annual level of heat consumption. An annual fee is agreed with the customer for the extent of service required and this includes the fuel, labour, maintenance, insurance, plant monitoring, water treatment, supervision, ash removal if coal or refuse fired, and electricity. Plant replacement is sometimes an option depending upon the age and type of the equipment.

Clearly, a detailed survey has to be carried out by the heat management company to assess not only the suitability and expected life and maintenance cost of the plant, but more particularly the quantity of fuel required to provide the necessary level of heating. The fuel consumption will be calculated precisely and will be the theoretical minimum required, consistent with the insulation levels, type of use, occupancy, and hours of heating required daily, weekly and for the agreed duration of the heating season. If necessary, the heat management company will install energy conservation equipment to ensure that the consumption levels are not exceeded.

Whilst the annual charge is fixed regardless of degree days, it is necessary to have a contract period of ten years to balance out the effect of mild and severe winters. If this is unacceptable to the customer, a fee is agreed which sets a target level for fuel consumption and an agreed variation for charges in degree days. Alternatively a saving scheme may be arranged, whereby both parties benefit if consumption is reduced below the target, and if the target fuel level is exceeded the client only has to pay a partial fee.

An adjustment formula is usually incorporated into the contract to provide for extra heating days or extended hours of operation during the heating season for the whole or part of the premises.

All these methods of fixed heating charges are designed to provide an incentive to the heat management company to be as efficient in plant operation and energy conservation as modern technology will allow, and to impose a financial penalty if it fails to achieve this. This system is undoubtedly one of the most effective in ensuring fuel consumption is maintained at optimum level.

Two-part Tariffs

This method of charging is used in cases where it is not possible to calculate the annual consumption level, e.g. with district heating schemes which have a wide variety of users, and with process steam users where the consumption is almost directly linked to output and not related to temperature.

The first part of the tariff is the standing charge which includes labour, insurance, maintenance, plant supervision, and plant monitoring. Any capital employed by the heat management company to improve efficiency, reliability, change from one fuel to a cheaper fuel, installation of conservation equipment, installation of modern boiler plant, heating systems etc. will also be included in the standing charge and will be paid off over the duration of the contract. A respectable heat management company will have tax shelter, and any assets required will be eligible for one hundred per cent first year allowances, the benefit of which will be reflected in the lower cost to the customer.

The second part of the tariff is the commodity or unit charge, which reflects the cost of the fuel used to generate the heat or steam at the optimum efficiency of the plant. It takes into account the average return temperature; steam raising plants usually have a contract which stipulates the extra or reduced cost if the customer varies his condensate return from the contracted level. It is considered that the higher the load factor on the plant, the greater will be the maintenance cost, and the higher the efficiency levels. Thus the unit charge reflects these two major cost items. Where electricity is the responsibility of the heat management company, this is usually reflected in the unit charge as well.

The selection of the two-part tariff means that actual heat or steam usage cannot be predicted, and thus it is equally impossible to calculate the level of supervision and administration work involved. Therefore, the required level of contribution to the heat management company's overheads and profit is spread over the standing charge and the unit charge on the same percentage basis. This ensures that the levels of charges are not increased to the customer in the event of his energy requirement being reduced.

Metered Charges

A heat management company may well choose to charge for the supply of heat, steam or electricity on a straight metered basis. However, unless the level of consumption can be assessed to within about ten per cent (when a fixed annual charge may possibly be arranged) then a 'minimum take' consumption figure is likely to be included in the contract. Failure to agree a minimum take figure could result in a reduced income and overhead contribution to the heat management company to a point where its labour costs and capital charges are not recovered; in practice, the level of charges is likely to be raised to cover for this possible risk.

The exception to this is a multi-customer scheme such as a very large district heating scheme where the reduced heat take by any one or even a number of major customers is unlikely to have any material effect on the overall plant output.

Escalation Clauses

All heat management contracts will include a 'rise and fall' clause, which will provide for changes in the cost of fuel, labour and materials; increases or decreases in these costs will be index-linked.

Fuel is usually the actual delivered price less appropriate rebate. The labour index may be a government-produced figure such as average earnings, or one related to an hourly rate. Materials index is used to cover increases in insurance, water treatment, repair and replacement costs etc. If electricity costs are included in the service, then these will also be linked to a suitable index—usually the average UK selling price of electricity. All indices will be stated in the contract and can readily be checked by the client to ensure that the level of charges made precisely reflects changes in costs on a strictly pro rata basis.

The heat management company may well have provided substantial capital plant; therefore, variations in the bank interest rate will cause a change in the charges.

The current Value Added Tax ruling states that supplies of fuel, heat, steam and electricity are zero rated, thus only the maintenance service work carries VAT.

Changes in government legislation which have a direct or indirect link to the cost of providing the service are usually subject to additional charges but referred to only in general terms within the contract.

Like the gas, water, and electricity supply utilities, heat management companies specify that they cannot accept any responsibility for consequential loss. However, it is usual for them to submit that their services and expertise will improve reliability and reduce the chance of plant downtime.

Energy Efficiency

Heat management is by definition the control and optimum utilisation of heat and steam provided to, and used in, offices, factory units, residential and commercial premises. It is therefore inevitable that heat management companies have built up a vast store of knowledge and experience in energy conservation. Indeed, energy conservation is virtually synonymous with heat management. The leading heat management companies are thus offering, as a part of their comprehensive package, the provision of sophisticated audit computer programmes, detailed surveys and, if required, a call out and maintenance service.

There are two ways in which heat management companies implement energy efficiency. In one a comprehensive survey of the customer's property is carried out. This determines exactly how much energy is required, for how long, at what temperature and even why it is actually needed. This information, together with full heat loss calculations and the capacity of existing heat emission surfaces plus any other energy uses such as refrigeration plant, lighting, lifts, etc will be used to evaluate and determine which of the various types of energy conservation and control devices are appropriate to each application. The company will then make recommendations indicating capital required (if any) and payback periods, and will install the appropriate equipment as instructed.

Alternatively, after a preliminary survey, the heat management company can take over the complete control and management of energy used in a building and in this way gain valuable practical operating experience with each installation. This will result in initial savings, while the longer term benefits of energy conservation equipment to suit the characteristics of the building can be properly assessed over a period of weeks. Again, recommendations will be made and equipment installed on instruction.

In both cases the heat management company can also provide the finance and recover the repayments out of the savings.

A range of complementary services are also available from heat management companies, including consultancy with particular reference to energy management and conservation. Maintenance-only services are usually offered, and have the advantage over maintenance-only companies in that a much wider variety of expertise and skills are available.

The enactment of the Energy Bill has given much needed impetus to combined heat and power, with applications in both the industrial and local authority field. The range of skills needed to evolve the concept, develop the market, design, finance, negotiate commercial sale and buy back terms with the electricity supply industry, coupled with the skilled operational demands of total energy systems are available from, and can often only be provided by, well-established and sound heat management companies.

Conclusions

Provided that the customer negotiates a fairly costed contract with a heat management company, then a strong argument can be made for acceptance of this bought-in service. The range of skills and technology available from a heat management company not only relieves the customer from being involved in a specialist activity outside his normal role, but does so at a cost which can often be less than he can achieve with his own staff. The services of a heat management company are particularly useful for conversion to coal which can be expensive and may require operating skills which the user does not command.

A PRACTICAL EXERCISE IN MONITORING AND TARGETING

Stan Smith

Manchester Steel Co Ltd was formed in 1974 by the Norwegian company Elkem Ltd to produce continuous cast billets from an Electric Arc Furnace. The main outlets for the steel were the British reinforcement market and the Rod Rolling and Wire Drawing Plant of Johnson & Nephew which was located adjacent to the plant.

Elkem Ltd purchased the Rod Rolling Plant of Johnson & Nephew in 1976 as part of their long term strategy to produce semi-finished steel products in the British market. The Rod Rolling Plant continued to trade as Johnson & Nephew (Manchester) Ltd for a period of five years but in 1981 the companies were combined to trade as Manchester Steel Co Ltd.

The 'Mini-Mill' has the capacity to produce 160,000 tonnes of continuous cast billet in 100 mm and 110 mm square sections and 220,000 tonnes of wire rod in the size range 5.5-13.5 mm diameter with a coil weight of 1.08 tonnes.

The plant layout is shown in Figure 10.1.

The Iron and Steel industry is a highly competitive and highly energy intensive industry. Success, particularly in the private sector, depends on close control of production and production costs, including energy costs.

The company had been routinely collecting information on energy use in its manufacturing processes for some time and, having been invited to take part in the Energy Efficiency Office's programme of pilot studies in monitoring and target setting in the steel sector, made a critical examination of the information it was gathering and the use being made of the information within management. This chapter explains how this company approached the target and monitoring exercise and how the information it accumulated was put to use to achieve energy savings.

The Target and Monitoring System

Setting up the monitoring system involved twelve steps. These are listed in Table 10.2 and commented on individually below.

Collection of Data

Most production units report on a weekly basis for management information and accounts. Energy consumption is usually taken as global figures—e.g. total gas metered, total electricity and total services (oxygen, water, fuel, oil etc.). This information can be roughly analysed into specific energy consumption related to production output.

Identification of Meters

All meters should be identified with details of areas they cover. Any major user areas which are not specifically covered by individual meters should be detected.

FIGURE 10.1: LAYOUT OF THE MANCHESTER STEEL COMPANY PLANT

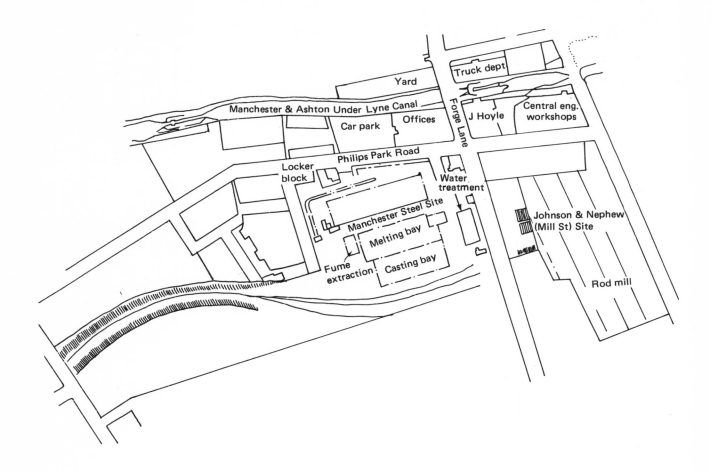

TABLE 10.2: SETTING UP THE TARGET AND MONITORING SYSTEM

1. Collection of all available energy/production data.
2. Identification of meters and areas covered.
3. Breakdown of areas into percentage of total energy used.
4. Selection of areas of high priority.
5. Detailed monitoring of selected areas over designated timescale.
6. Analysis of monitored information.
7. Breakdown of priority areas into specific parts.
8. Preparation of standard operating procedures for each specific part.
9. Setting of agreed targets based upon historical data and best practice.
10. Preparation of monitoring system using agreed targets.
11. Setting up the reporting structure.
12. Initiating regular reviews.

Percentage Energy Usage of Designated Areas

Using the existing meters, measurements should be taken over a period of two weeks. This detailed information should then be tabulated to indicate major energy users.

Selection of High Priority Areas

Using the information obtained above, priority areas need to be selected for full targeting and monitoring. Meters should be installed where necessary, in all defined high usage areas to enable detailed readings to be taken. It is not necessary to measure each individual item of plant.

Detailed Monitoring of Selected Areas

After the designated meters have been installed it is necessary to decide on a frequency for meter readings—e.g. daily, weekly. Monitoring sheets are prepared for collection of the information by the responsible plant personnel. The information is then transferred to a weekly report which is forwarded to the designated Energy Controller. This monitoring procedure is carried out over a period of four to six weeks.

Analysis of Monitored Information

The analysis of the information thus obtained will determine which areas require a detailed break-down of plant operation. The Energy Controller will carry out this work in conjunction with the respective Plant Managers.

Breakdown of Priority Areas

Most of the plant can be broken down into specific operating practices which have a direct bearing on energy consumption. This stage requires the input from all personnel involved with the plant operation.

Standard Operating Procedures

After the standard operating procedures have been finalised it is necessary to carry out a training programme to ensure that plant personnel are fully conversant with the reasons for change.

Setting of Agreed Targets

Using historical data, when available, in conjunction with the information obtained during the initial four to six week monitoring period, targets are set for each segment of the plant operation. It is important that the targets are agreed with the responsible Plant Manager and are realistic.

Monitoring System

The monitoring system must be robust to ensure consistency. The information gathered is discussed by each Plant Manager during his morning production meeting, with action points referred to each Section Manager. After the initial trial period it will become obvious to each Plant Manager that the monitoring periods should be reduced to individual shifts or periods of production. In the case of the rod mill the monitoring was carried out on a shift basis. On the steel plant the monitoring was carried out on a cast by cast basis. Reaction time was therefore reduced and action points completed quickly.

The storing and analysis of the data becomes an increasing problem. This company had microcomputers available which enabled software to be prepared for carrying out the mundane data collection and presentation.

The CUSUM method of analysis was used with great success on the Rod Mill Furnace monitoring. It is essential that any plant or operating procedure modifications should have readily identifiable savings.

The Reporting Structure

It is vital that the project has the correct level of management involvement. This ensures that priorities are correctly identified and progress is maintained.

At Manchester Steel the Project Group was made up as follows:

—Energy Controller   —Chief Engineer
—Plant Controller   —Plant Manager

Each Plant Manager delegates areas of responsibility within his plant which involves Section Managers and Shift Managers. The Shift Manager is the catalyst within his own shift. Reporting cycles were established as follows:

—Plant Manager : daily
—Energy Controller : weekly
—Works Manager : monthly

Project Group Meetings

After a period of assessment including the readjustment of targets it became apparent that a co-ordinating meeting should be held at three monthly intervals to analyse targets and monitoring procedures.

Summary

It is essential that, in the initial stages, specific items of plant, which have the most effective return, are chosen for the scheme. If too many items are chosen this dilutes the reaction time. When the system is operating automatically further assessment should be carried out to extend the areas of monitoring.

The Energy Audit

A detailed energy audit was carried out for the total site. The results are indicated in Tables 10.3-10.7. From the information obtained it was decided to carry out a target and monitoring exercise on the following items:

—Steelplant
  —Electric Arc Furnace 33 kV
  —Horizontal Ladle Preheater
  —Emergency Preheater
  —Fume Extraction Plant

—Rod Mill
  —Billet Reheating Furnace
  —Electricity Consumption
  —Central Workshop

Analysing Energy Consumption and Operating Procedures

The Electric Arc Furnace

A detailed analysis of the standard operating procedures was undertaken with special reference being made to those areas which would directly affect energy consumption. These were as follows:

  —power on time for each basket
  —time taken to charge each basket
  —matching of furnace performance to casting performance
  —oxygen usage—oxy fuel burner and oxy lance
  —delay costs—energy related.

Large discrepancies were found to occur from shift to shift. Discussions were held with the Shift Manager and key operatives on each shift to quantify the energy losses associated with their practices. The correct practices were explained and detailed monitoring sheets were prepared for each cast. Targets were prepared using 'best practice' and included on the monitoring sheets.

Horizontal Ladle Preheater

It was found that there was no standard operation for carrying out ladle preheating. Each shift had different methods for achieving the final preheat temperature; the final temperature also varied a great deal. Standard operating procedures were prepared to quantify the preheat time and final temperatures. A gas meter was installed to monitor performance.

Emergency Preheater

Again no standard operating procedures were available. The tundishes in this area were poorly designed for heat retention and were held at no predetermined temperature.
   A ceramic fibre lined lid was manufactured and standard operating procedures introduced. A gas meter was installed to monitor performance.

Fume Extraction Plant

A detailed inspection revealed that although the extraction fan was required to be running at all times during the production operation the fan was left running until Sunday 0600 hours due to previous problems with dust fallout. It was decided to switch off the fan progressively earlier on Saturday and monitor the operational problems. The extraction fan motors are 2 x 150 hp.

TABLE 10.3: ANNUAL FUEL CONSUMPTION

Fuel Type	Total Units	Therms	Percentage
Electricity	86,122,100 kWh	2,936,764	44.1
Natural Gas	3,686,438 Therms	3,686,438	55.4
Fuel Oil	90,245 Litres	32,578	0.5
Total		6,655,780	

TABLE 10.4: ANNUAL BREAKDOWN OF NATURAL GAS

Area	Therms	Percentage
Steel Plant	787,251	21.4
Steel Plant Amenities	17,704	0.5
Welfare	2,700	⟨0.1
Rod Mill	2,812,225	76.3
Rod Mill Stores Heating	6,434	0.2
Rod Mill Laboratory	10,046	0.3
Forge Lane Offices DHW	4,784	0.1
Forge Lane Old Canteen	1,508	⟨0.1
Central Engineering	43,786	1.2
Total	3,686,438	

TABLE 10.5: ANNUAL BREAKDOWN OF ELECTRICITY

Area	kWh	Percentage
Arc Furnace (33 kV)	60,319,000	70.0
Steel Plant (6.6 kV) [1]	6,780,500	7.9
Rod Mill (6.6 kV)	19,022,600	22.1
Total	86,122,100	

(1) A recent change in the method of calculating electricity bills has meant that the previous distinction between the rod mill and the steel plant has been lost. The figures shown are based on the same ratio of steel plant to rod mill consumption being applicable for the last three months as was true for the previous nine months.

TABLE 10.6: 6.6 kV SUPPLY—BREAKDOWN OF USAGE

6.6 kV Supply Area	Average	Percentage of Total
Water Treatment 1	300	} 38.4
Water Treatment 2	230	
Fume Extraction	300	21.7
Continuous Caster	150	10.7
Furnace Auxiliaries	150	10.7
Compressor Plant	150	10.7
Lighting	50	3.6
Oxygen Plant	50	3.6

Figures do not add due to rounding.

TABLE 10.7: ANNUAL BREAKDOWN OF SERVICE ENERGY

Item	Fuel	Consumption Therms	Percentage
Steel Plant—LPHW Boiler [1]	Natural Gas	14,700	8
Rod Mill Steam Boiler and Radiant Heating [1] [2]	Natural Gas	51,610	28
Steel Plant Amenities LPHW Boiler	Natural Gas	17,704	10
Welfare LPHW Boiler	Natural Gas	2,700	1
Rod Mill Stores Radiant Heaters	Natural Gas	6,434	3
Rod Mill Laboratory LPHW Boiler	Natural Gas	10,046	5
Forge Lane Offices DHW	Natural Gas	4,784	2
Forge Lane—Old Canteen	Natural Gas	1,508	1
Central Engineering—Air Heaters	Natural Gas	43,786	24
Forge Lane Offices LPHW	35 Sec Fuel Oil	32,578	18
Total		185,850	

(1) Assessed (all other consumptions are metered).
(2) Includes all year heating from steam boiler for lubricating oil system.

TABLE 10.8: ROD MILL—BILLET FURNACE WEEKLY ENERGY FIGURES

	DAY	MONDAY			TUESDAY			WEDNESDAY			THURSDAY			FRIDAY		
	SHIFT	6-2	2-10	10-6	6-2	2-10	10-6	6-2	2-10	10-6	6-2	2-10	10-6	6-2	2-10	10-6
*	PREHEAT ZONE Cu Ft	0				0			0			0				0
*	TONNAGE ZONE Cu Ft															0
*	SOAK ZONE 1 Cu Ft															0
*	SOAK ZONE 2 Cu Ft															0
	TOTAL Cu Ft															0
	THERMS															0
*	BILLETS PUSHED TONNES	0				0			0			0				0
*	THERMS/TONNE	0				0			0			0				0
*	COIL TONNES	0				0			0			0				0
	YIELD %	0				0			0			0				0
*	TIME AVAILABLE	0		480	480	0	480	480	0	480	480	0	480	480		0
*	TIME ACTUAL	0				0			0			0				0
*	AVAILABILITY	0				0			0			0				0
	ELECTRICITY KWH															
	ELECTRICITY T/T															
	ELECTRICITY KW/T															
	THERMS/TONNE															
	GAS METER READING (Cu Ft x 100)															

TABLE 10.9: ROD MILL ENERGY ANALYSES

```
ANALYSES OF GAS (THERMS) LIGHT UP & HOLD PERIODS (TARGET 6-2 (T50(2-10 850)
=============================================================
```

WEEK NO.	MONDAY 6-2	TUESDAY 2-10	WEDNESDAY 2-10	THURSDAY 2-10	TOTAL
2	7879	918	688	1013	·****
3	308	1034	856	961	3762
4	491	982	361	· 971	3406
5	2424	1201	1123	1086	5841
6	2204	1160	1045	1107	5538

```
***********************************************************************
```

```
ANALYSES OF SPECIFIC GAS USAGE (THERMS/TONNE) (TARGET DAILY <16.0 WEEKLY <19.0)
================================================
```

WEEK NO.	MONDAY	TUESDAY	WEDNESDAY	THURSDAY	FRIDAY	WEEKLY
2	43.12	17.24	18.29	17.52	17.36	21.00
3	29.83	16.23	17.69	18.85	17.73	19.33
4	23.03	19.04	17.98	20.88	17.83	20.40
5	24.94	20.88	18.67	17.16	16.75	19.36
6	22.75	19.13	17.77	16.68	15.41	18.12

```
***********************************************************************
```

```
ANALYSES OF ELECTRICITY CONSUMPTION (TARGET DAILY <3.8 WEEKLY <4.0)
================================================
```

WK.NO.	TUESDAY1400HRS KWH/T	TH/T	WEDNESDAY1400HRS KWH/T	TH/T	THURSDAY1400HRS KWH/T	TH/T	FRIDAY2200HRS KWH/T	TH/T	WEEKLY KWH/T	TH/T
2	416.6	*.**	109.6	3.7	121.5	4.10	164.3	5.60	193.3	6.60
3	180.1	6.10	138.4	4.7	126.4	4.30	100.6	3.40	131.0	4.47
4	144.4	4.90	154.3	5.3	155.0	5.30	157.7	5.40	152.5	5.21
5	166.8	5.70	135.2	4.6	103.3	3.50	116.1	4.00	128.7	4.39
6	153.2	5.20	134.6	4.6	113.4	3.90	95.5	3.30	121.4	4.14

```
***********************************************************************
```

```
ANALYSES OF PERCENTAGE GAS USAGE/ZONE (TARGET PH2. G5:TZ. 45:S1. 11:S2. 9)
================================================
```

WEEK NO.	P.H.ZONE	TON.ZONE	SOAK 1	SOAK 2
2	22.07	61.42	11.13	5.38
3	26.44	61.05	8.13	4.36
4	26.75	62.02	7.34	3.90
5	26.92	59.69	9.02	4.37
6	30.81	59.09	6.91	3.19

```
***********************************************************************
```

```
CUSUM ENERGY DATA ANALYSES
================================
```

WEEK NO.	T/PUSHED	ACTUAL THERMS	CALC. THERMS	SAVINGS THERMS	SAVINGS POUNDS
2	2533	58835	50746	-8088	-2588
3	2516	54600	51826	-2774	- 987
4	2322	50922	47704	-3218	-1029
5	2580	52350	51021	-1329	- 425
6	3094	57475	58387	+ 912	+ 292

```
***********************************************************************
```

Billet Reheating Furnace

The existing consumption records were based upon a total site gas meter reading which gave very little information relating to the performance of the furnace alone. As an initial starting point gas consumption was monitored for each day for each furnace heating zone and a gas meter installed to measure the service gas used for the remainder of the plant. The gas consumption figures were analysed for 1982-83 and these were used as the base line for 1984.

Typical target and monitoring sheets are shown in Tables 10.8 and 10.9.

Electricity Consumption in the Rod Mill

The electricity meters were read weekly for management information. This gave a total consumption for electricity for the rod mill, central workshop and main office block. Due to the multiplicity of meters involved to determine individual area usage it was decided to monitor the main meter only and record daily consumption.

Central Workshop and General Service Fuel

No previous monitoring was carried out in these areas. In the initial stage it was decided to monitor all service fuel consumption on an individual area basis and concentrate target and monitoring for the central workshop gas usage. Existing meters were available for all the service areas.

Establishing Targets

Based on a careful appreciation of the data collected a series of target energy savings were established and agreed with staff. These are set out in Table 10.10.

TABLE 10.10: AGREED TARGET SAVINGS

Area	Savings (percentage)
Electric Arc Furnace	0.5
Steel Plant Natural Gas	13.0
Rod Mill Natural Gas	1.0
Rod Mill Electricity	3.0
Services	1.0

Results Achieved

Electric Arc Furnace

During the first two months of 1984 detailed analyses and monitoring of each cast's operating procedures were carried out. Best practices were noted and standard operating procedures with targets were prepared for various groups of steel qualities.

Training seminars were held with all shift managers and foremen to ensure that the new operating procedures were understood. The shift managers were made responsible for carrying out individual shift training programmes to ensure that all key operatives were aware of the reasons for implementing the new procedures.

At the daily production meeting target and monitoring were included on the agenda and the previous day's energy performance was discussed. Using the targets set, corrective action points were noted and progressed with the individual shift managers. An analysis of the 1983 results was used to prepare the CUSUM data. The best line was found by the least squares method—see Chapter 1. The equation for the line was found to be:

Electricity consumption (33 kV) = 437.9 x tonnes hot metal + 194,130
(Correlation coefficient = 0.97)

The CUSUM results are shown in Figure 10.11. As can be seen from the graph a rapid deterioration in savings occurred after week 32 when an increased transformer capacity was installed. Further detailed work is in progress to assess the overall performance for production and energy. The effect of target and monitoring was therefore only considered over the weeks 1-29, as the parameters were the same as 1983.

The saving indicated from weeks 1-29 is 1 million kWh or 2.7 per cent, equivalent to £25,000.

Rod Mill Billet Furnace

An analysis of gas consumption and tonnes pushed through the furnace was carried out for 1982-83.

The best line least squares method produced the following equation for the line:

Gas consumption (therms) = 18.30 x tonnes pushed + 6,000
(Correlation coefficient = 0.96)

This equation was used as the calculated value in the CUSUM table.

Major modifications were subsequently carried out to the furnace. In 1983 the roof, sidewall and sloping roof were coated with ceramic fibre tiles. In week 5 of 1984, proportions of power to each zone were modified. This work was completed in week 13. In week 8, 1984, billet temperature recorders were fitted. The main drive loads were monitored and billet temperatures reduced to match main drive loads. In week 22, 1984 the end wall was coated with ceramic tiles and in week 31, 1984 the charge refractory door was replaced with chain curtain and woven heat resistant material.

The results for 1984 were recorded using the CUSUM method and are shown in Figure 10.12.

From this graph the annual savings were shown to be 160,000 therms or 5.1 per cent, equivalent to £49,600.

The performance of the furnace had been monitored over the period 1982-84. The effectiveness of the modifications to reduce energy consumption was calculated again using the best line least squares method:

—1982: gas consumption (therms) 18.89 (tonnes pushed) + 5,050
—1983: gas consumption (therms) 14.62 (tonnes pushed) + 16,370
—1984: gas consumption (therms) 14.02 (tonnes pushed) + 15,150

In 1982 the rod mill was operated on a 15-shift production basis and changed to a 10-shift operation in February 1983.

The major difference between the shift systems was holding the furnace at a predetermined temperature for the shifts where no production took place. This resulted in energy consumption of 2,700 therms per week.

FIGURE 10.11: STEELPLANT ELECTRIC ARC FURNACE

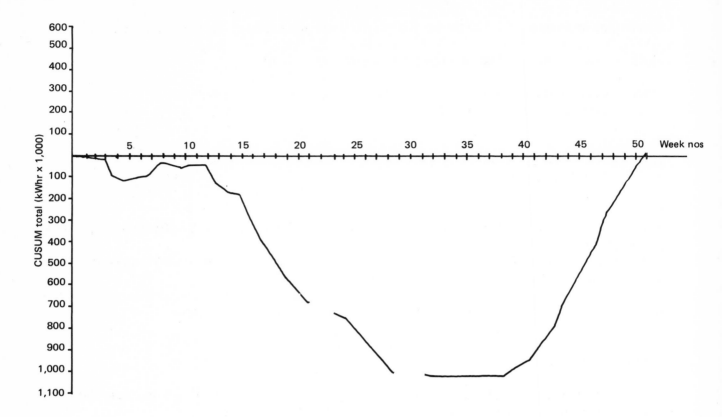

FIGURE 10.12: ROD MILL BILLET FURNACE 'CUSUM' GRAPH

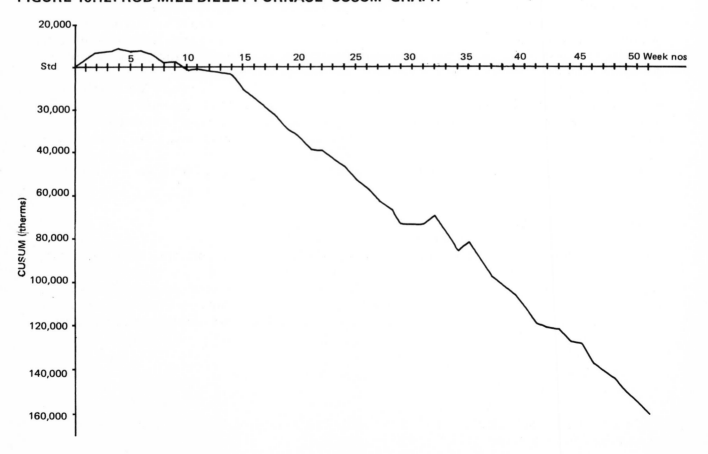

Cost Effectiveness

Based upon an average of 3,300 tonnes pushed per week and gas costs of £0.31 per therm

1982   Cost = (18.89 (3,300) + 5,050) 0.31 = £20,890

1983   Cost = (14.62 (3,300) + 16,370) 0.31   =   £20,031
        Less cost of hold periods (2,700 x 0.31)   =   £   837
                                              Total   =   £19,194

1984   Cost = (14.02 (3,300) + 15,150) 0.31   =   £19,039
        Less cost of hold periods (2,700 x 0.31)   =   £   837
                                              Total   =   £18,202

The ceramic tiles installed in August 1982 resulted in annual savings of £81,408 giving a payback period of three months.

The introduction of target monitoring and the modifications previously indicated resulted in an annual saving of £49,600.

The monitoring of energy consumption is part of the morning production meeting and readings have been increased from weekly to each shift to improve action response time.

Central Engineering Workshop

The workshop has a dual function; nearly half of the floor area is utilised as a machine shop and assembly area and the other half for storage, mainly of refractory bricks.

The heating is provided by two dravo gas-fired air circulating boilers. The original control was based upon the thermostatic control of the outlet air temperature from the boiler and switched on/off manually as required.

Modifications were carried out to enable the boilers to be thermostatically controlled by measuring the workshop air temperature. A time clock was also fitted to each unit to ensure that the units were turned off during the non-occupation period. These modifications were carried out during the summer of 1984.

Gas consumption had been monitored weekly from July 1983. The cumulative gas consumption was plotted against each month (Figure 10.13) for 1983 and 1984. Comparing the second six months of 1984 with the corresponding period in the previous year, savings of 6,400 therms were reported, equivalent to 41.3 per cent of previous consumption and representing a financial saving of £2,073.

Allowing for degree day compensation, Figure 10.14 shows savings of 34.4 per cent (£1,726).

Forge Lane Office Block

The consumption of the central heating system fired by fuel oil was monitored from June 1983.

Calculations were made based upon monthly oil consumption divided by degree days for the appropriate month. Cumulative totals and monthly totals were graphed for each month. This graph for 1983 was used as the target for 1984—see Figure 10.15. The cumulative totals were compared for the period August to December for each year, as this reflects the temperature difference for each year:

-1983: indicator 27.47
-1984: indicator 29.49

This represents a comparable loss of 7.35 per cent. This method of analysis is being used for 1985 to set targets for monitoring.

FIGURE 10.13: CENTRAL ENGINEERING WORKSHOP

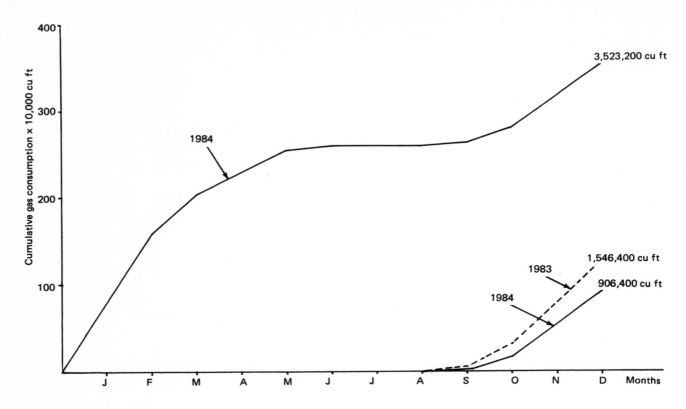

FIGURE 10.14: CENTRAL ENGINEERING WORKSHOP GAS HEATING

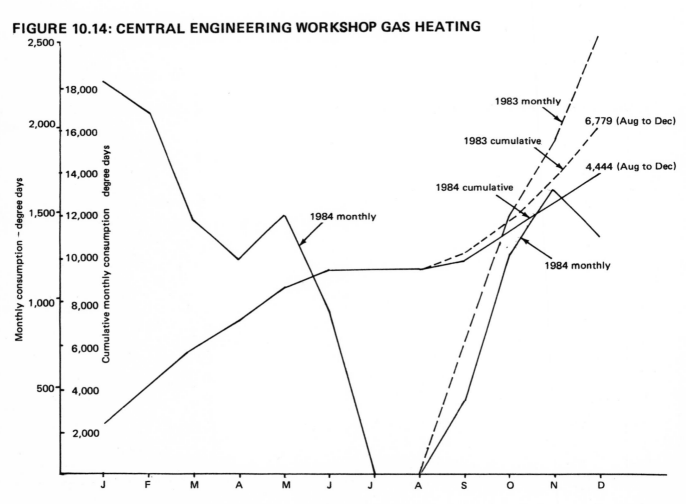

FIGURE 10.15: FORGE LANE OFFICE BLOCK OIL CENTRAL HEATING

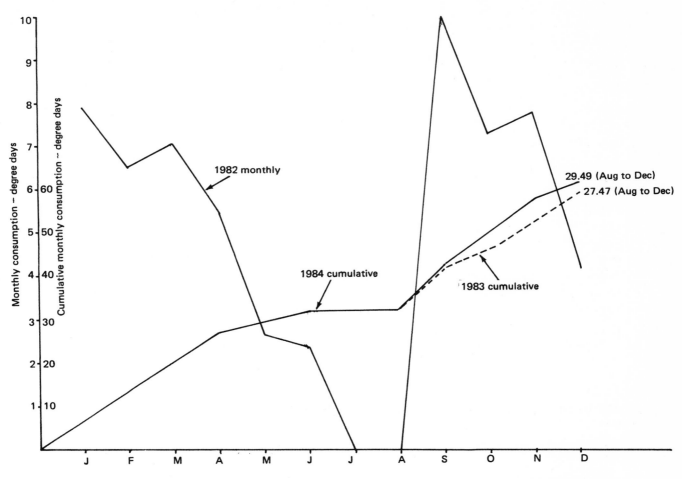

FIGURE 10.16: STEELPLANT LOCKER BLOCK GAS HEATING (EXCLUDING SHOWERS)

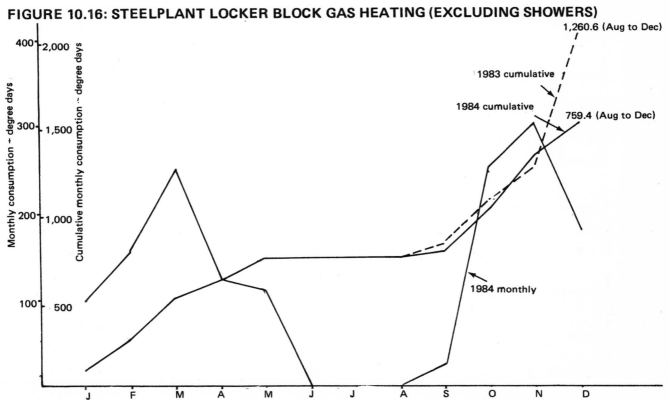

Steelplant Locker Block

The locker block contains showers which are used by all steelplant personnel. The gas boiler provides hot water for showers as well as central heating for the block. The steel plant is manned 24 hours per day for five consecutive days. Single shifts are also worked Saturday and Sunday mornings.

Checks on the system during the 1983 summer period revealed a set consumption for shower hot water. This figure was used as the standard for each month. The difference between consumption and standard consumption was accepted to be the central heating load.

Cumulative totals and monthly totals divided by degree days were graphed for each month and these are shown in Figure 10.16. The periods August-December were monitored for each year. The large increase in December 1983 was noted and steps taken to ensure the system was only used when required over the Christmas holiday period in 1984. Savings of close to 40 per cent were identified.

Rod Mill 6.6 kV Electricity Consumption

An analysis of 1983 results revealed a poor correlation between billet tonnes pushed and 6.6 kV consumption. The equation of the line was found to be:

y = 52.24 (tonnes pushed) + 225,482
Correlation coefficient 0.427, standard deviation 47,541.

It was decided to compare 1984 results using a graph of tonnes pushed/6.6 kV consumption using best line ± 1 standard deviation and to analyse results above and below 1 standard deviation—see Figure 10.17.

FIGURE 10.17: ROD MILL BILLET—TONNES PUSHED

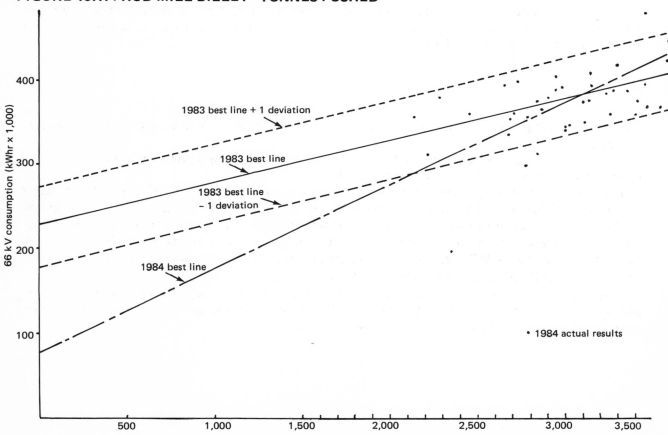

Studies were carried out to determine start up procedures and plant shutdown requirements during extended delays. These were modified to minimise energy consumption, bearing in mind the risk factor of increased production delays due to start up faults extended past anticipated start up time.

These modifications proved to be effective when one considers the analyses of 1984 results. The best line produced an equation y = 98.54 (tonnes pushed) + 78,818, with a correlation coefficient of 0.7333, and a standard deviation of 44,491. The average kWh per pushed tonne was reduced to 124.64 in 1984 from 125.68 in 1983 and represented a saving of £4,500.

The target for 1985 will be based upon 1984 best line $\pm$ 0.5 standard deviation.

Steel Plant Natural Gas

Due to the developments carried out throughout the year on the horizontal ladle preheaters, the furnace oxygen/natural gas burners and methods of operation, it was extremely difficult to set realistic targets using 1983 as base. Periodic problems were also experienced with the accuracy and calibration of the natural gas meters.

The problem with the individual meters highlighted the need to carry out site energy balance monitoring based upon the collated information and the Gas Board meter which covers total consumption for the steelplant site.

Target and monitoring in this area was therefore not considered appropriate although modifications to the actual time the ladles are on preheat reduced the time by eight hours per day.

Steel Plant 6.6 kV Electricity

Examination of the fume extraction plant, which incorporates 2 x 150 hp induced draft fans, revealed that the fans were stopped 18 hours after production had stopped, due to problems previously experienced with dust holding in the hoppers of the baghouse. The time was gradually reduced to two hours after production stopped without any detriment to the baghouse operation. This resulted in a saving of £4,000 per annum. Future work will be carried out in 1985 to reduce water treatment plant consumption.

Conclusions

Target monitoring was the catalyst to concentrate the company's resources to minimise energy consumption.

If one has to highlight a single part of the target and monitoring scheme which gave the most benefit it must be the CUSUM method of analysis.

The company's existing energy monitoring policy was based upon accounting type management information. This was obviously retrospective and greatly limited the reaction of individuals.

Using a combination of targets and CUSUM greatly reduced the reaction time needed to change operational practices to minimise energy consumption. Any changes in practice could immediately be analysed by the CUSUM method to determine energy savings. Comparing this result with any detrimental effect on production would give the net savings.

Target and monitoring itself will not realise savings without the constant input of all key personnel associated with energy consumption. The responsible manager must react to any change from target gain or loss to ensure effective targets are maintained.

TABLE 10.18: TARGET AND ACTUAL SAVINGS ACHIEVED

Item	Area	Target (percentage)	Actual (percentage)	Period Cost Saving	Annual Cost Saving
1	Electric Arc Furnace	0.5	2.7	25,000	25,000
2	Steel Plant Natural Gas	13.0	—	—	—
3	Rod Mill Natural Gas	1.0	5.1	49,600	49,600
4	Rod Mill Electricity	3.0	1.0	4,500	4,500
			Sub Total	79,100	79,100
5	Services	1.0	—	—	—
5.1	Central Workshops	—	34.4	1,700	3,400
5.2	Office Block	—	(7.35)	(500)	(1,000)
5.3	Locker Block SP	—	39.7	400	800
		1.0	12.0	1,600	3,200
6	Fume Extraction	—	—	4,000	4,000
Total				84,700	86,300

During the period of target and monitoring it became increasingly obvious that the two main factors limiting accurate monitoring were:

—human errors in reading meters and the actual timing of meter readings
—reliability of the meters in operation

To ensure reliability any future meter installed will have a digital readout, in place of the dial, and there will also be included an automatic recorder indicating time and consumption. This would allow for validation of total site consumption by summating all sub meters and comparing with master meter reading.

Acknowledgements

BNF Metals Technology Centre
National Industrial Efficiency Service Ltd.